THE BIONEERS

The BIONEERS

Declarations of Interdependence

Kenny Ausubel

Chelsea Green Publishing Company
White River Junction, Vermont

Printed in the United States

First printing, October 2001
04 03 02 01   1 2 3 4 5
Printed on acid-free, recycled paper

Library of Congress Cataloging-in-Publication Data
Ausubel, Ken.
   [Restoring the earth]
   The bioneers : declarations of interdependence / Kenny Ausubel.
      p. cm.
   Originally published: Restoring the earth. Tiburon, Calif. : HJ Kramer, c1997. With new pref.
   Includes bibliographical references and index.
   ISBN 1-890132-76-4 (acid-free paper)
   1. Environmental degradation. 2. Environmental protection. 3. Human ecology. 4. Biology. 5. Sustainable development.   I. Title.
   GE140 .A87 2001
   333.95--dc21   2001047170

Chelsea Green Publishing Company
Post Office Box 428
White River Junction, VT 05001
(800) 639-4099
www.chelseagreen.com

We did not come into this world. We came out of it,
like buds out of branches and butterflies out of cocoons.
We are a natural product of this earth, and if we turn out
to be intelligent beings, then it can only be because we are
fruits of an intelligent earth, which is nourished
in turn by an intelligent system of energy.

Lyall Watson, *Gifts of Things Unknown*

CONTENTS

PREFACE TO THE
SECOND EDITION

LATELY, I'VE BEEN LIVING WITH A WEIRD SENSE of cognitive dissonance. On the one hand, storming around us is the destruction of life and disruption of the basic biological systems on which all life depends. It's frightening, depressing, and unspeakably painful. A masterful public relations juggernaut casts nets of disinformation, ensnaring masses of people in delusions of complacency, as though the war against the Earth is perfectly natural, inevitable, and even necessary to maintain our "standard of living."

Yet at the same time, a vibrant culture of innovative solutions is being born out of this cataclysm and it is spreading rapidly around the world. Extraordinary human creativity focused on problem solving is exploding the mythology of despair. Over and over, it's the story of how an individual can make a difference. The answers percolate up from the profound wisdom of the natural world. This is where the Bioneers live. Being steeped in this fertile brew of remedies, I feel as much cause for hope as for horror.

When millions of computers worldwide rolled over into year

2000, the brittle global grid of accidents waiting to happen didn't. Hundreds of billions of dollars and swarms of feverish digital plumbers averted the predicted apocalypse shadowing the centralized technocratic plundering of the world. The pre-millennium was a yawner. Industrial society rolled on inexorably. Prophecies dropped like flies. Techno-utopian civilization triumphed. Unless, of course, you turned away from the media spectacle and actually looked around.

Just weeks earlier, tens of thousands of activists gathering in Seattle had conjured a tsunami. Waves of resistance washed over the carefully crafted globalization agenda and exposed its abuses of the environment and human rights. Seattle cleaned the World Trade Organization's clock, setting back its timetable by years. Cornerstones of the design to commodify the world crumbled. Deep fissures among member nations split wide open the most basic assumptions about who's in charge and just what kind of future the world's people want. Teamsters and Turtles joined hands to reboot democracy, unexpectedly scrambling the circuits of what had seemed an immutable program.

Meanwhile, reports surfaced among farmers in Iowa that cows breaking into fields of genetically modified corn held up their noses and turned tail for more familiar pastures. Sometimes there *is* a place for animal testing! People in Europe and Japan seemed to agree with the cows, and before long Frankenfoods were stuck in the craw of global trade. When major markets like Europe and Japan fall out, the globalization business plan craters. What had seemed a done deal was now up for grabs.

And on New Year's Eve 2000 in France, half the trees went down in a freak tempest, yet another of those 500-year storms that have lined up since the 1990s with the compressed frequency of overbooked jet planes. Among the places hardest hit were the grandfather groves at the Palace of Versailles. A French national treasure, many of these trees were planted during the reign of Louis XIV. Best remembered for his remark "L'état, c'est moi"—"I am the state"—the Sun King lacked the foresight also to claim domain over the weather, whose shifting patterns are by now unequivocally linked with the bonfire of fossil fuels stoking up the planetary heat.

People tuning in to the Weather Channel have watched it turn into an action-adventure movie.

But the alarmists were wrong about climate change: They underestimated it. Scientific consensus now indicates that we can expect climate change to progress from two to ten times faster this century than in the last. At the same time, the impending biodiversity crash threatens to extinguish a fifth to two-thirds of the world's species in this next hundred years. The consequences are unimaginable, from shredding the very fabric of evolutionary resilience just when it is needed most, to making this planet a mighty lonely and impoverished place. Public health is already showing serious strains from the 80,000 or so synthetic chemicals now suffusing the most intimate tissues of the world, leaving mother's breast milk so toxic that it could not legally be sold on store shelves in many countries.

By the time the actual calendrical millennium punched in, it was living up to its promise of biblical times. If there is a single message in all this turmoil, it's that the biological world is a permeable membrane, infinitely interconnected. We have made a basic systems error in believing that we are somehow separate from the natural world. As human beings, we are one with the Earth. It's time to come home.

As physicist and author Fritjof Capra suggested at the 1999 Bioneers Conference, "Concern with the environment will no longer be one of many single issues. It will move to the center of the stage. It will become the context of our lives, our businesses, our politics."

The good news is that for the most part the solutions are present. Where we don't know exactly what to do, we have a pretty good idea what direction to head. Janine Benyus, author of *Biomimicry*, puts it eloquently: "The people learning about biology are reaching back to ancestors that are 3.8 billion years old. In that time, life has learned to do amazing things: fly, circumnavigate the globe, live at the top of mountains and bottom of the ocean, lasso solar energy, light up the night, make miracle materials like our skin, horns, hair, brains. Life has done everything that we want to do but without guzzling fossil fuels, polluting the planet, mortgaging its future. What better models could there possibly be? We can decide as a culture to listen to life, to echo what we hear. We are surrounded by geniuses. Learning from them will take only stillness on our part, a quieting of the voices of our own cleverness. Into this quiet will come a cacophony of earthly sounds, a symphony of good sense. The real benefit of a life of biomimicry is that we begin to feel a part of, rather than apart from, this genius that surrounds us."[1]

One of the great beauties of biology is that its facts are also our metaphors. In the coming pages you'll read about the ant-fungus combination, a wonder of mutual aid so elegant that it has endured uninterrupted for tens of millions of years. As biologist E. O. Wilson relates, their cooperation is so complete that it's impossible to tell if the ants captured the fungus to serve their needs, or the other way around. Recent research has revealed an even grander complexity of symbiosis than we previously imagined.

Biologists were puzzled about how the ants could keep their precious fungus safe from pathogens, an especially dire threat to such a monoculture. It turns out that their caverns are *not* free from pathogens. In fact, the fungus is very vulnerable to a devastating mold found only in ants' nests. To keep the mold at bay, the ants long ago made a discovery that would send the stock of any pharmaceutical company soaring.

Scientists now know that the ants have domesticated at least four kinds of fungus, which are often cultivated nearby to one another. When one fungus is threatened by the mold, which happens within two years among 60 percent of colonies, the ants bring in another species that provides a bacterium lethal to the deadly mold, and is also a fertilizer for the fungus. That same bacterium is the source of over half the antibiotics used in human medicine. "Some Alexander Fleming of an ant discovered antibiotics millions of years before people did," wrote Nicholas Wade.[2]

Ants also invented agriculture before people did, by 50 million years, and they are finessing two challenges beyond the reach of present industrial society. They are continuously growing a monocultural crop without disaster, and they are dispensing antibiotics so frugally as to avoid provoking antibiotic resistance. Commented one of the biologists, "It may be one of the best studied symbioses in biology, but it is a sad reflection of how little we know in general."

These are the true biotechnologies. They exemplify a kind of magical realism, a tantalizing glimpse into how solutions residing in nature vastly surpass our prior conceptions of what is possible.

Innovators around the world are tapping into this kind of ancient biological intelligence to devise practical solutions to many of our most pressing crises. The Bioneers are on the forefront of this movement to reweave the sacred web of life and our proper place within it.

THE BIONEERS: REVOLUTION FROM THE HEART OF NATURE

The Bioneers you'll read about in these pages are biological pioneers who have peered deep into the heart of living systems to see what we can learn from four billion years of evolution. What are nature's operating instructions? How would nature do it? What they are unearthing is nothing less than a revolution from the heart of nature. In many cases, their knowledge is prefigured by ancient indigenous traditions, many being validated by modern research. The applications are spreading with encouraging speed.

Since I first wrote about architect William McDonough, his work has continued to define the Next Industrial Revolution. His firms are now working with the Ford Motor Company to redesign the famous River Rouge plant in Michigan where Henry Ford created the first vertically integrated automotive manufacturing facility. Bill is leading a $2 billion project intended to become what Ford's Chairman, William Clay Ford, Jr., describes as "the icon of the Next Industrial Revolution." Bill's design includes restorative native landscapes as well as the largest habitat roof in the country— a 454,000-square-foot green invitation to birds. He's advising on the manufacturing of automobiles, too. Bill has penetrated the highest circles of corporate design with comparable projects, from Nike and the Gap to Ciba-Geigy and Herman Miller. He was rightly named a "Hero of the Planet" by *Time* magazine, and was awarded the nation's highest environmental award, the Presidential Award for Sustainable Development, in 1996. He continues to promote putting the filters in designers' heads, not on pipes or smokestacks.

The kinds of changes championed by Bill McDonough and other Bioneers are already visible in many sectors. New York State has taken the lead in making commercial and residential buildings more efficient and environmentally friendly by adopting "green building credits" against state income taxes. Spain has initiated the use of olive oil and olive biomass wastes to power major utilities. The number of giant power-generating windmills in that country has doubled every year since 1995, to fourteen hundred, a number expected to reach nine thousand by 2010. Along with Germany and Denmark, Spain is providing state funds to offset the cost differential for consumers between renewable energy sources and the market price. Across the globe during the 1990s, wind power installations grew at a rate of 26 percent a year, solar photovoltaics at 17 percent annually. While these technologies still produce only one percent of the world's energy, double-digit growth rates can change the picture quickly. The alternative energy industry may well mimic the birth of the oil industry 100 years ago in its vertiginous expansion.

As Paul Hawken and Hunter and Amory Lovins describe in their landmark book *Natural Capitalism*, many large companies, governments, and global institutions are finding that ecological practices can cut costs by half, quadruple profits, and create more new jobs in the process. The jobs are knowledge-intensive rather than labor-intensive. "Less stuff, more people" is the mantra, in keeping with the state of the world. Given the one billion unemployed and underemployed people living in dire poverty, economic justice is a precondition for environmental well-being.

Bioneer Jason Clay has continued to demonstrate the economic viability of converting the world's major commodity crops to sustainable production. Working with shrimp aquaculture, he has shown large producers how they can simultaneously diminish environmental damage and improve profit margins. These Better Management Practices (BMPs) pay for themselves in two to three years. One of the lessons is that these practices have a social dimension. In Latin America, he has found that shrimp producers

can have four times the yields and profits when workers ar
bonuses and incentives to better manage feed use, which itself is a
major source of pollution. Investors now see these BMP-based
screens as a way to reduce financial risks. The marriage of ecology,
economy, and social justice pays off. Jason is now gearing up for
model demonstration projects with oil palm, soy, cocoa, orange
juice, and sugar.

John Todd's unique work with "living machines" has also continued to blossom in entrepreneurial ventures. He has been working to convert the South Burlington, Vermont, living machine from
sewage treatment to an agro-eco-industrial park. Using wastes from
a brewery, it will act as a farm to grow vegetables, fish, and flowers,
and to produce new crops for pet foods, thus serving as a model of
bioenterprise. In Maryland, he is working with a large food producer to use his unique floating pond restorers to turn chicken
wastes into beneficial food webs, instead of suffocating Chesapeake
Bay.

John's richly simulated solar ecosystems are now successfully
treating "wastes"—i.e. cycling nutrients—from wineries and brew-
eries in California. A dazzling living machine graces the new Environmental Studies Center at Oberlin College in Ohio, a state-
of-the-art facility designed by Bill McDonough in collaboration
with the Lovinses. Treating all the facility's sewage, the system runs
purified water outside to nourish the Midwestern prairie and
woodlands. The center serves as a focal point for educating people
about what is already possible in ecological building. One of the
elders of biomimicry, John currently teaches as a Distinguished
Professor of Ecological Design at the University of Vermont, a singular program in the country, bringing these ideas to new generations of designers.

Agriculture is also on the cusp of transformation, marked by a
steady transition to organic and sustainable practices. The organic
food market is racing ahead at 20 percent growth a year in the
United States, highlighting the ardent public hunger for a safe and
nutritious food supply. Over 40 percent of food production in Aus-
tria is currently organic. Pending legislation in Britain would push
organic production to 30 percent by 2010; demand is so great there

that 80 percent of organic foods must currently be imported. The city of Munich, Germany pays farmers in the watershed that supplies its drinking water to farm organically. The future is organic.

Farmer, philosopher, and farm policy specialist Fred Kirschenmann, whom you will meet in these pages, has gone to seed. With good reason. Though few people realize it, organic foods are not grown from organic seeds. An arcane statute permits U.S. farmers to use nonorganic strains if a sufficient supply of organic seeds is not available. This Catch-22 has discouraged the growth of an organic seed sector.

Witnessing the stealth-food takeover of U.S. farming by genetically modified seeds, Fred helped launch a seminal counteroffensive through his Northern Plains Sustainable Agriculture Society. With several other groups, they are working to ramp up organic seed production for farmers. Similar to the model used by Vandana Shiva and the Indian seed movement (described here as well), the plan is to generate a green necklace of loosely connected regional organic seed cooperatives. But because virtually all current agricultural seeds are hybrid varieties designed to grow with petrochemical inputs, Fred has also brought diversity into the equation, reviving traditional varieties and breeding new ones specifically adapted to organic cultivation and local ecologies.

The virtue of organic diversity is not merely aesthetic. A recent experiment in China found that planting two varieties of rice instead of a uniform monoculture doubled the yield and virtually eliminated its most devastating disease—without the use of any chemicals and without costing more. This was not backyard stuff— the experiment took place across 100,000 acres on tens of thousands of commercial farms. Similar practices are being tried with barley in Europe and coffee in Colombia.

Fred recently became director of the Leopold Center for Sustainable Agriculture in Iowa, where he can disseminate these kinds of models and ideas more widely (www.leopold.iastate.edu). The center's mission is to promote on-the-ground research that will mitigate the damage of current agricultural practices and promote alternatives. Early efforts on stream restoration around farms were so effective that both the Environmental Protection Agency and

U.S. Department of Agriculture have become involved to create a national demonstration site. Fred also continues to manage his large biodynamic farm.

Vandana Shiva's activities have also continued to bear fruit. Despite the ongoing onslaught of corporate agribusiness ravaging traditional Indian farmers, community seed cooperatives are thriving in ever-widening circles of grassroots diversity. Indian activists launched a successful challenge against the patent on the neem tree granted to a giant multinational corporation. An icon of Indian botanical culture, the neem is considered part of the people's collective heritage dating back thousands of years. Vandana herself has stepped even more prominently onto the world stage to redress such critical issues of biopiracy and the theft of traditional knowledge. Following the Seattle uprising, she was invited to the World Economic Forum in Davos, Switzerland, where she had the ear of globalization advocates, who now are listening.

Steven King's Shaman Pharmaceuticals found out the hard way that drug development is a jungle. A series of setbacks, including prohibitive costs and incessant regulatory delays, forced the enterprise to reconstitute itself as an herb company. Given fierce disagreement in the botanical community about the value of a single "active ingredient" adapted for a drug distinct from the complex chemical synergy of the whole plant, it may be for the best. Perhaps this is what the plants wanted all along.

Organic greenhouse master Anna Edey compiled her wealth of knowledge into a book, *Solviva: How to Grow $500,000 on an Acre and Peace on Earth*. Jennifer Greene, following her recuperation from a serious car accident, is back in the flow of water restoration. Don Hammer retired after thirty years as a Johnny Appleseed of constructed wetlands, leaving a glistening green legacy of purified water and habitat for critters. Josh Mailman is still seeding visionary enterprises with "adventure" capital funds. Kat Harrison has taken her teaching on the road, bringing enchanted stories of living ethnobotany to audiences from the Massachusetts College of Pharmacy to the Boston Museum of Science.

Finally, John Perkins's Dream Change Coalition has spread its vision of an earth-honoring dream beyond the Amazon and Andes with shamanic workshops in Africa, Europe, the Himalayas, Siberia, and Central America. Each year the group sponsors a "Gathering of Shamans" at the Omega Institute in upstate New York. John also put together a book of rare interviews with his Shuar allies, *Spirit of the Shuar: Wisdom from the Last Unconquered People of the Amazon*. He is now organizing to change the dream in schools across the United States.

The Bioneers Conference has grown and prospered. The eight hundred attendees in 1997 swelled to over twenty-six hundred in 2000. The voices of the Bioneers now reach millions of listeners through "Bioneers: Revolution from the Heart of Nature," an international radio series, and through articles and conference excerpts distributed in the print media. The Bioneers Web site (www.bioneers.org) provides access to solutions and strategies for people working for restoration around the world.

Bioneers is not a spectator sport. Virtually all who attend the conference are actively engaged in restoration. If the Bioneers community is any kind of indicator, we're likely to see a resurgence of activism—people acting on their values—as problems intensify and solutions become more widely known. It's up to all of us to help this earth-honoring dream come true.

A DECLARATION OF INTER-DEPENDENCE

Early in the 20th century, naturalist Aldo Leopold shared some of his greatest insights about nature and human culture: "All ethics so far evolved rest on a single premise: that the individual is a member of a community of interdependent parts. His instincts prompt him to compete for his place in that community, but his ethics prompt him to cooperate. A land ethic then reflects a conviction of individual responsibility for the health of the land.

"A thing is right when it tends to preserve the integrity, stability, and beauty of the biotic community. It is wrong when it tends otherwise. The last word in ignorance is the man who says of an animal or plant, 'What good is it?' If the land mechanism as a whole is good, then every part is good, whether we understand it or not.

If the biota, in the course of aeons, has built something we like but don't understand, then who but a fool would discard seemingly useless parts? To keep every cog and wheel is the first precaution of intelligent tinkering."[3]

At this critical moment in the fate of the Earth, we have an extraordinary opportunity to work with nature to heal nature, and ourselves with it. What we are learning above all is the intricate interconnectedness of living systems. Listening to nature is providing us with the antidotes, grounded in a deep capacity for self-repair. The overarching enterprise of restoration is one of healing.

James Lovelock, who developed the Gaia Hypothesis that the Earth is an intelligent, self-organizing kind of superorganism, recently reflected on how he now sees this vision evolving. "Gaia theory sees the biota and the rocks, the air and the oceans as existing as a tightly coupled entity. Gaia's evolution is a single process and not several processes studied in different buildings of universities. It has a profound significance for biology. It affects even Darwin's great vision, for it may no longer be sufficient to say that organisms that leave the most progeny will succeed. It will be necessary to add the proviso that they can do so only so long as they do not adversely affect the environment.

"A geophysical system always begins with the action of a single organism. If this action happens to be locally beneficial to the environment, then it can spread until eventually a *global altruism* results. The reverse is also true, and any species that affects the environment unfavorably is doomed, but life goes on. Gaia works from an act of an individual organism that develops into global altruism."[4]

Gaia's biological logic now appears to be playing out through the human species, and not a moment too soon. Mounting numbers of people are awakening to the reality that what is good for Gaia is good for us.

COMING HOME TO THE EARTH

After an aching divorce from a long-term marriage, I was sure I would never wed again. Then I met Nina. We fell madly in love.

Four years later, we were still going strong. I reconsidered. Why let an old wound hobble the fullness of our love together?

We didn't want an off-the-shelf ceremony, so we decided to create something more personal, to make it ours. Instead of exchanging traditional vows, we would trade "wows." We agreed not to share them until the ceremony itself, in the intimate company of many of our dearest friends and relations. As the day grew near, I was seized by nameless, formless terror.

I spilled out my quaking fear to a close friend over lunch. He flashed a coyote grin and reared back with laughter. Any truly transformational experience is preceded by dread, he counseled me.

To work, the changes demanded of us today must be transformational. There are many deep wounds to heal, not least those of the human spirit, which is as capable of creation as of destruction. We know from restoring ecosystems that they rebound with vitality and strength when given encouragement. You get what you give.

A wedding of sorts beckons us today: the consecration of our renewed wows with the natural world. The Bioneers herald for this new century and millennium a Declaration of Interdependence, a celebration of the sacred recognition that we are the trees, the land, the water, the whales, the redwoods, the mushrooms, the microbes.

How very fortunate we are to get to come home at last.

NOTES
1. Janine Benyus, speech at the 1997 Bioneers conference.
2. Nicholas Wade, "For Leaf-Cutter Ants, Farm Life Isn't So Simple," *New York Times*, Aug. 3, 1999.
3. From "The Land Ethic," passim, in Aldo Leopold, *A Sand County Almanac* (New York: Oxford University Press, 1949).
4. James Lovelock, "Gaia as Seen through the Atmosphere: The Earth as a Living Organism," in *The Biosphere and Noosphere Reader*, Paul R. Samson and David Pitt, eds. (London: Routledge, 1999), p.119.

PREFACE TO THE
FIRST EDITION

As I stood behind the cameraman, I watched the sun glint off the bright red corn seeds held gently in the upraised hand of the Native American farmer. He had unearthed these scarlet seeds in a little clay pot embedded in the mud wall of his adobe home. Showing them around the community, he found only a couple of elders who still remembered what they were. They marveled at the sacred red corn of San Juan Pueblo, which no one had grown in forty years. Tears rose in the farmer's voice. This humble man planted the sacred seeds, and was now harvesting a spiritual homecoming in the ancient tribal community in the Rocky Mountains of New Mexico.

In these seeds lived not only the genetic legacy of countless generations of Pueblo farmers, but also the imprint of their hands. These crinkly scarlet seeds glowed with the songs, prayers, and gritty growing secrets of all the farmers who had come before, a tenuous heritage that must be applied to be preserved. Their rediscovery would lead to a historic revival of native agriculture at the pueblo.

In that farmer's hands, I saw the profound interconnectedness of human beings with the Earth. A slender thread binds the weave tight in the intricate, mysterious fabric of life, and yet we have been blindly unraveling the tapestry of creation. As we strain the limits of the natural world, we can no longer escape the knowledge that ecological collapse has been the hidden force behind the downfall of many major civilizations. Biology is indeed destiny.

I thought I was at San Juan Pueblo to make a film, but it turned out I was there to start a seed company. I went on to cofound Seeds of Change, a mission-driven venture devoted to conserving the world's ark of agricultural seeds by commercializing them in a market partnership with backyard gardeners. Shortly after I started the company, I initiated the Bioneers Conference, galvanizing a budding but disparate culture around the "biological model" of interconnectedness. I saw that we needed to go beyond a human-centered Declaration of Independence to a biological Declaration of Interdependence. The time had come to unite nature, culture, and spirit in service of the restoration of Earth and our relationship to the web of life.

The one simple, overarching fact to understand about the state of today's environment is that *all the basic life-support systems on Earth are in serious and accelerating decline.* At the same time, our human population is rising precipitously. While statisticians may argue about the particular status of a given species, toxin, or regional ecosystem, we cannot escape the fact that the biology of our planet is groaning under the accrued weight of human abuse.

Previous civilizations—the Mayan, the Mesopotamian, the Mediterranean—have similarly ignored the natural limits of their time and met their demise.[1] In today's truly global and relentlessly efficient high-technology world, our actions hold the capacity to topple civilization on a planetary scale.

As author Paul Hawken has noted, *sustainability* (the idea that we must not diminish natural systems beyond their capability to replenish at comparable levels) is simply the midpoint between destruction and restoration.[2] Because we live in such a dramatically depleted world today, it is not enough to seek merely to sustain it.

Sustainable systems may be a reasonable long-term goal in a regenerated world, but for now we need to tip the scales toward *restoration* to regain the subtle equilibrium that is the nature of nature.

To restore the Earth, inspired individuals have set about learning the mysteries of renewal from nature itself, where there resides that most astonishing capacity for regeneration. These *bioneers*—biological pioneers working to restore the Earth—have been growing in knowledge and numbers, and they are having an increasingly positive effect throughout global society. Just as the tumbling of the Berlin Wall was both unforeseen and necessary, the biocide we have been conducting against Earth seems destined to fade away into a pile of souvenir rubble. Any other future is almost unimaginable, and certainly not viable.

Since I first visited San Juan Pueblo in 1985 and witnessed the return of the sacred red corn, the community has revived its ancient farming tradition and built a modern food-processing facility. In a culture where "food is life," the restoration of "agri-culture" is both a material and spiritual act. Those humble corn seeds booted up a 40,000-year-old farming legacy, which the Pueblo believes will carry it successfully into the next century, reestablishing the ecological balance between the tribe and the Earth in a sacred partnership.

This book is intended to scan the emerging horizon for the relief map of the positive future that beckons us, when as human beings we will learn to weave our lives in harmony with the natural world and embrace the spirit of the Earth.

Kenny Ausubel
Spring 1997

ACKNOWLEDGMENTS

I WOULD LIKE TO THANK THE BIONEERS WHOSE WORK AND vision are reflected here. Since I have begun seeing the world through their eyes, life has never looked the same to me. The glint of magic and mystery shines from within the natural world, and the bioneers have helped me learn to see it more brightly.

I especially want to thank J. P. Harpignies, who assisted me greatly with the compilation of the factual chapter introductions, and who also lent his wise and astute perspectives throughout. He has also been an indefatigable associate producer of the Bioneers Conference almost since the beginning.

As always, I give deep thanks to my partner and wife Nina Simons, who has tirelessly critiqued this manuscript at every phase, using her keen eye and true heart, and who has been a singular partner in coproducing the Bioneers Conference since its inception seven years ago.

Special thanks go to Josh Mailman, the godfather of the Bioneers Conference who shared the birth of the vision.

Thanks to the many people who have helped produce the Bioneers Conference, including Suzanne Jamison, Barbara Whitestone, Chris Shea, Maggi Banner, Sebia Hawkins, Claire Greensfelder, and Adam and Adelina Raskin.

I also extend our heartfelt gratitude to our sponsors, including Greg Steltenpohl and Stephen Williamson and Odwalla, Tom Van Dyck and Progressive Asset Management, Ed Alstat and the Eclectic Institute, Aubrey Hampton and Susan Hussey and Aubrey Organics, Carolyn Mugar and Farm Aid, Jay Harris and Mother Jones, Jennifer Barclay and Marc Wallach and Blue Fish, Robert Henrikson and Earthrise Trading Company, Paul Dolan and George Rose and Fetzer Vineyards, Wild Oats Community Markets, Todd Koons and TKO Farms, Patrice Wynne and Eric Jost and Gaia Bookstore, Marc Kasky and Fort Mason Center, Joshua Grossman and Dancing Tree, Lisa Conte and Shaman Pharmaceuticals, John Goodman and Arrowhead Mills, Bill St. John and Nutraceutix, Inc., New Hope Communications, Rodale Institute, Bioremediation Services, John Roulac and Harmonious Technologies, Tami Simon and Sounds True, Randy Hayes and Rainforest Action Network, Horst Rechelbacher and Aveda, Maria Gallardin and TUC Radio, Alan and Marian Hunt-Badiner and the Roy A. Hunt Foundation, Bob Weir and the Further Foundation, Rockwood Fund, Humani-Tees Foundation, Jeffrey Bronfman, Susannah Schroll, Susanna Dakin, Glenda Anderson, André Carothers, Nick Morgan, Carol Brouillet, and all the Bioneers community who have come together to celebrate the restoration of Earth.

Thanks to my agent and friend Steven J. Schmidt for sharing the vision and staying with the project across shifting sands.

I would also like to thank Paul Hawken, who has been an inspiration and beacon to me for many years. He brings both great insight and heart to the work.

Thanks to Chronicle Books for permission to reprint the poetry of Francisco X. Alarcón.

Thanks to Jesse Ausubel for his thoughtful reading of the work, and to Anne Ausubel for instilling in me an early love of the outdoors and writing.

Very special thanks to Susannah Schroll for her etheric house-keeping, and to Jeffrey Bronfman for his unerring courage and kindness.

Special thanks to my business shaman Ben Dover.

In the eventful four years since this book's initial publication, many people have contributed mightily to the ongoing work of Bioneers. Two merit very special mention. Susanna Dakin, artist and visionary philanthropist, made a five-year grant that allowed Bioneers to thrive and grow beyond anyone's expectation. Her generosity, vision, and heart are the invisible ink behind many of our endeavors. And Ginny McGinn, who joined Bioneers in 1998 as managing director, has been the force of nature assuring that this handful of an enterprise is implemented with exceptional skill, integrity, and grace.

Finally, I dedicate this book to my daughter Mona in hopes that the world will be a better place for her and her generation as a result of the work of the bioneers.

"All My Relations"

THE SIGNS ARE EVERYWHERE, WHEN YOU KNOW WHERE TO LOOK. LIKE GRASS sprouting through the pavement, the force of life is reclaiming the Earth and infiltrating culture the moment the machines go silent and their guards fall asleep. Awareness of the inevitability of global restoration is spreading steadily among giant corporations, government institutions, grassroots activists, and individuals and communities around the world.

Restoring the Earth is destined to be the central enterprise of the years ahead, and leading the effort is a growing movement of bioneers, biological pioneers who are using nature to heal nature. As environmental restoration emerges as a major global industry, the bioneers are acting as the pilot fish guiding the dynamic transition to a future environment of hope. Though it is not widely known, the bioneers have devised practical solutions for virtually all our crucial environmental problems, and the problems are truly grave.

BROTHER, CAN YOU SPARE A PARADIGM?

Try as we might, human beings have been unable to repeal the law of gravity or regulate the weather. In that sense, biology is indeed destiny, as the ultimate fate of our species rests upon our ability to live within the limits of the natural world.

As we witness the widespread degradation of the basic life-support systems upon which human and all biological life depends, we are realizing that Earth itself is stronger than we can imagine and will certainly survive our folly. But human beings as a species may not prevail, and we are

1

permanently destroying countless other species in a cascade of extinction bearing the unprecedented mark of the human hand. Many scientific experts tell us that we have only the decade of the 1990s to launch a fundamental reversal of the devastating damage we have visited upon Earth.[1]

Environmental conditions are also precipitating mounting political crises, as journalist Robert Kaplan disturbingly illustrated in his article "The Coming Anarchy." Characterizing the environment as a "hostile power," he wrote, "For a while the media will continue to ascribe riots and other violent upheavals abroad mainly to ethnic and religious conflict. But as these conflicts multiply, it will become apparent that something else is afoot, making more and more places like Nigeria, India, and Brazil ungovernable. It is time to understand 'the environment' for what it is: *the national-security issue of the early twenty-first century.* The political and strategic impact of surging populations, spreading disease, deforestation and soil erosion, water depletion, air pollution, and, possibly, rising sea levels in critical, overcrowded regions like the Nile Delta and Bangladesh—developments that will prompt mass migrations and, in turn, incite group conflicts—will be the core foreign-policy challenge from which most others will ultimately emanate, arousing the public and uniting assorted interests left over from the Cold War."[2]

How have our "advanced" science and technology led our society to the brink of this most unnatural disaster? What is our place in the natural world? How do we set about restoring Earth, our home?

In fact, practical solutions do exist for almost all our environmental problems. Dedicated bands of bioneers have been persistent in creating and demonstrating innovative models. They span fields as divergent as restorative agriculture, ecological design, power generation, water purification, natural medicine, biodiversity conservation, economic renewal, cultural revival, and spiritual politics. They have produced successful prototypes and shown their effectiveness. These working models hold the keys to our survival and can be refined, replicated, and rapidly spread around the world. The optimistic bioneer vision is already empowering individuals, communities, groups, and companies to act as a primary force in the fabric of restoration.

The bioneers herald a coming Age of Biology, one in which as human beings we ground our lives within the limits of the natural world. Imitation is the sincerest form of flattery, and we are learning to emulate natural systems to fashion benign technologies and just forms of social organization. Above all, the bioneers call for us to grow a culture that connects with the sanctity of all life through the heart of nature.

THE SUPERB ART OF RELATIONSHIP

Biology is not rocket science. It is far more complex and subtle than the physics of sending metal to the moon. Ecology is the superb art of relationship in the fantastically elaborate web of life. As geneticist David T. Suzuki observes, "In fact, so exquisitely orchestrated is this global system of flowing energy and matter that the biosphere as a whole seems at times to have the self-perpetuating qualities that we normally associate only with intact living organisms. The entire Earth, within its extraordinary capacity (within limits) to regulate and maintain its own global ecological balance, acts as if it were a single great 'superorganism' of planetary proportions."[3] This concept, elaborated as the Gaia Hypothesis by scientists James Lovelock and Lynn Margulis, views the very Earth as a whole living system that is self-regulating.[4]

Margulis, a revolutionary microbiologist who has spent her life studying bacteria, happily underscores the fact that we live in a microbial world. Our bodies may be little more than convenient hosts for the three-and-a-half-billion-year-old microbial community. Microbes, which make up at least 80 percent of the Earth's biomass, are now believed to have been the dominant player in creating and arranging the rocks, seas, oil, soils, gases, metals, and minerals of the Earth's surface.[5] Without these organisms, all life would cease because decomposition would no longer occur and we would be buried in a matter of weeks under a heap of eternal biomass. Perhaps we should learn some respect for this invisible microbial polyverse, which cohabits our own universe and generously harbors us as guests.

The mystery of life and living organisms is probably so intricate and profound as to be ultimately unknowable by human beings. Esteemed Harvard biologist Edward O. Wilson paints this remarkable tableau of biological relationships among leafcutter ants and fungi: "The leafcutter colony is a single organism. . . . The superorganism's brain is the entire colony. . . .

Through a unique step in evolution taken millions of years ago, the ants captured a fungus, incorporated it into the superorganism, and so gained the power to digest leaves. Or perhaps the relation is the other way around. Perhaps the fungus captured the ants and employed them as a mobile extension to take leaves into the moist underground chambers. . . . In any case, the two now own each other and will never pull apart. The ant-fungus combination is one of evolution's master clockworks, tireless, repetitive and precise, more complicated than any human invention and unimaginably old."[6]

It is this sort of model of symbiosis that infuses the work of the bioneers. Their solutions mimic natural forms and systems with "biomorphic" (imitating biological forms) technologies based on nature and containing live intelligence. They embody complex dynamic systems characterized by constant change. Bioneer systems have a pulse. They are alive and sovereign.

THE BIONEER VISION

Who are the bioneers? They are a diverse culture of improbable players — field biologists, anthropologists, entrepreneurs, gardeners, artists, public servants, poets, ecologists, architects, botanists, and activists. This book surveys a partial cross-section of fields where their endeavors are percolating to the surface. Even the expansive diversity of their broad explorations reveals a common ground of values and codes.

The first bioneer notion is *kinship*. At the molecular core of life, from microbes to mammals, organisms share far more in common than not. Only a tiny fraction of DNA separates chimp from hominid. The modern sciences of evolutionary biology, genetics, and ecology have revealed an intimate inner connection among the stuff of all life forms from the inside out.

Nor is it solely competition that makes the natural world go round, as some Darwinists have asserted. In contrast to Darwin's view of "survival of the fittest," biologist Lynn Margulis sees evolution as a family affair. "We consider naive the early Darwinian view of 'Nature red in tooth and claw.' Now we see ourselves as the products of cellular cooperation—of cells built

4

up from other cells. Partnerships between cells once foreign and even enemies to each other are at the very root of our being."[7]

The bioneers agree on *diversity* as a core condition of life. Studies have found that ecosystems rich with diversity are strong and resilient in the face of environmental crisis and change, whereas ecosystems impoverished of diversity are weak and vulnerable. Diversity lies at the heart of evolution. Without it, organisms cannot successfully adapt to change, the one constant in nature. The predator that extinguishes its prey must also perish. What nature teaches is *interdependence*.

Another bioneer principle is the idea of a *food web*. In nature, there is no "waste." Everything is something's "lunch" of food or energy. There are no by-products, only products. How would we change our behavior if we truly comprehended the spiral cycles of biological life?

Community is also a keystone of biology, say the bioneers. Although ecological systems may have especially noteworthy individual players, the life of the system is conducted by the cooperative symphony, not the soloist. Community is a primary biological unit, a higher octave of cooperation.

The greatest communality among the bioneers is a *spiritual connection to nature*. Nature is their wellspring, their teacher, their inspiration, their comfort. They express a profound humility in the unknowable face of the grand mystery of nature. There they find nourishment, a spiritual source, and they say that the essence of any practical solution for restoring the Earth is a change of heart. What spirituality means to them is quite simply a reverence for life. Without it, no "solutions" can repair our environmental damage.

MUTUAL AID

The damage we have inflicted on planet Earth is indeed severe. There are few easy fixes. The situation demands massive action by millions of people, a level of mobilization seldom seen except in times of war. Indeed, we have been conducting a war against nature, and we have maintained a state of deep denial about the depth of biocidal harm we are doing. How do we find our way through the thicket? How do we discover our place on planet Earth?

5

The Russian anarchist philosopher Prince Peter Kropotkin was a naturalist who probed the mystery of the biological nature of human values. "Two aspects of animal life impressed me most during the journeys which I made in my youth in Eastern Siberia and Northern Manchuria," wrote the scholar in his celebrated 1916 pamphlet *Mutual Aid*. "One of them was the extreme severity of the struggle for existence which most species of animals have to carry on against an inclement Nature; the enormous destruction of life which periodically results from natural agencies; and the consequent paucity of life over the vast territory which fell under my observation. And the other was, that even in those few spots where animal life teemed in abundance, I failed to find—although I was eagerly looking for it—that bitter struggle for the means of existence, *among animals belonging to the same species*, which was considered by most Darwinists (though not always by Darwin himself) as the dominant characteristic of struggle for life, and the main factor of evolution. . . .

". . . . Whenever I saw animal life in abundance . . . I saw Mutual Aid and Mutual Support carried on to an extent which made me suspect in it a feature of the greatest importance for the maintenance of life, the preservation of each species, and its further evolution. . . . The idea that Mutual Aid represents in evolution an important *progressive* element . . . begins to be recognized by biologists."[8]

Edward O. Wilson has proposed a parallel concept to Kropotkin's benevolent vision of Mutual Aid in "biophilia," the affinity that life has for life. Perhaps in biophilia lies redemption, imprinted in our cellular capacity to experience the empathy that all living beings innately can share with one another.

The original bioneers, the world's indigenous peoples, have long viewed their principal duty as learning to live in a relationship of mutual aid with Mother Earth. As tribal chief Black Elk once said, "It is the story of all life that is holy and is good to tell, and of us two-leggeds sharing in it with the four-leggeds and the wings of the air and all the green things. For these are the children of one mother and their father is spirit."[9]

In fact, we know very little about the natural world. The so-called "balance of nature" may well be illusory, as a certain degree of chaos might be embedded in nature. Recent scientific studies suggest that fluctuation is the norm and that equilibrium is at most a transitory state. The term *chaos* as it is used by scientists implies the idea of order within disorder. "Turbulent, chaotic systems may look as if they behave randomly and with total disorder. But scientists say they exhibit an underlying, long-term order that is neither wholly deterministic nor wholly random."[10] The issue becomes one of seeing the patterns and aligning with them. Life is change.

To make the successful transition to a restorative, just, and healthy world, we need to understand ourselves as biological creatures at one with the diversity of all life. When we can truly see this unity and interdependence, say the bioneers, we will find nature to be forgiving, generous, and resilient.

With direct action, the bioneers are improving the environment by changing the world. What they seek is nothing less than a new covenant with nature, the "reenchantment" of the Earth. In restoring the Earth, we can restore our spirit. Native American peoples offer a humble prayer for the life of the world that says simply, "All My Relations." Expressed in innumerable indigenous languages around the world, this elegant phrase signifies our kinship with all life.

GATHERING THE BIONEERS

During many years of personal interest in environmental affairs, I have had the good fortune to meet a series of remarkable people who have rekindled my deepest hopes for the restoration of the Earth. These bioneers, as I eventually came to call them, are impassioned individuals who each hold an astonishing window on biological restoration. Their ideas and practices are pragmatic and original, brimming with creativity and ingenuity. They are visionaries with both feet on the ground.

I was surprised that few of the bioneers knew of one another's work. I felt dismayed that the world generally had little or no idea of their existence. Here were visionaries offering eminently practical solutions and insights into many of our most dire environmental threats, yet often they were languishing in isolation, sometimes struggling to survive just till morning. They represented valiant efforts to bring about environmental improvement and social justice.

My own work, which revolved mostly around plant medicine, filmmaking, and writing, evolved in 1989 into the cofounding of Seeds of

Change. Sitting around one day with my friend and business associate Josh Mailman, an activist organizer, investor, and philanthropist profiled in this book, I was dreaming about somehow bringing these visionary people together. Josh expressed his enthusiasm for a conference and offered to help fund it. Thus was born the Bioneers Conference in 1990.

The conferences, which I coproduce with my partner and wife Nina Simons, originally focused on two principal areas: biodiversity and bioremediation (the use of natural treatment systems for decontaminating soil and water). After three years, the event outgrew Santa Fe, and we moved it to San Francisco, where it has flourished since. Many progressive, mission-driven companies now support the event, as well as private foundations and individual donors. The program also expanded into other areas, including culture, spirituality, and feminine vision.

The Bioneers Conference is solution-oriented, inspiring people with a vibrant sense of optimism and empowerment. The event helps create "action groups" among the bioneers themselves. It also gives the public direct access to the "experts," a situation which offers people real tools and guidance to implement restorative practices in their own lives, communities, and companies. As anthropologist Margaret Mead once observed, "Never doubt that a small group of thoughtful committed citizens can change the world. Indeed, it's the only thing that ever has."

Since the bioneers started gathering in 1990, we have witnessed the community grow and seed movements and industries. People around the world are coming together around these commonly held ideas and a spirit of restoration. The bioneer vision is already especially strong in the personal health movement, where "green medicine" appeals to very large numbers of people and is entering the mainstream; in some sectors of the business community, where these kinds of ideas make fiscal good sense; and in the public sphere, where personal commitment to environmental health is surprisingly strong.

This book covers eight key areas of bioneer effort, but it is not intended as an encyclopedia. Rather it provides a window into crucial changes that are well underway on many of the most critical issues we must address to restore the environment.

The changes needed today are so vast that thousands and millions of people are called upon to take personal responsibility to act on behalf of the environment. There is neither time nor reason to await the dragging heels of business-as-usual. Perhaps the most important contribution of the bioneers is that they show innumerable ways that we can all begin to make a tangible difference in our personal lives and in our larger communities and political domains.

Seeing through the eyes of the bioneers will likely change the way you see the world. Clearly we cannot proceed much farther down our current path. It is not a question of *whether* to change, but of *how*. The bioneers open up a series of parallel universes that are often so hidden to most of us that they will probably come as a wondrous surprise in their sheer idiosyncrasy.

We face an extraordinary opportunity for the restoration of our home, the Earth, and with it our spirit. As the Age of Biology beckons, it is not only the threat of destruction that propels us forward. Instead, the bioneers show, it is the inspiration and magic of life that will finally prevail. As we gaze in the mirror of history, we find ourselves facing the inescapable knowledge that we inhabit one planet, indivisible and imbued with intelligence and spirit.

It is my hope that this book will help spread awareness of some visionary yet eminently practicable solutions and bring greater support for these endeavors and others like them. We need the vision, the political will, and economic means to implement them. Like magnetic north, the bioneer vision is drawing society in a powerful direction. We don't have much time, but clearly we have the models and the creativity. If people have created the problem, people can also create the solution. And—if only we pay attention—we have all around us the greatest teacher there is: nature.

CHAPTER ONE

Cleopatra's Bathwater

WHEN BIOLOGISTS SPEAK ABOUT "CLEOPATRA'S BATHWATER," WHAT THEY
are playfully illustrating is that the Earth is mainly a closed loop. What's
here today was generally here yesterday. So that cup of tea you're drinking
could once have been Cleopatra's bathwater! But if Cleopatra were to
bathe in the Earth's waters today, her skin would crawl and palace heads
would roll. Our booming human population is massively disturbing the
planetary waters, polluting and straining the ecology of water everywhere.

Water and oxygen define the blue-green Earth as a "water planet," in
Jacques Cousteau's lovely image. The human body, which is 97 percent
water, can survive longer without food than without this vital liquid. Yet
freshwater available for human use constitutes less than half a percent of
all water on the planet, and represents an increasingly scarce and threatened resource. Even the once seemingly endless seas are beleaguered.

As water wizard Peter Warshall has written, "Life depends *in toto* on
water's constancy. (Technically, this is one aspect of an organism's homeo-
stasis.) The ability of water to absorb large amounts of energy buffers
photosynthesis in cytoplasm and the transfer of oxygen in animal blood
from chaotic flux; moderates the Earth's climate by using oceans and lakes
for heat storage; eases seasonal change and our bodies' adaptation to it by
slowing, without shocks, the change of weather; and protects plants like
cacti from boiling under desert skies. Most of all, water's specific heat, heat
of vaporization, and heat of fusion give life its ability to maintain itself in
hard times. Without these molecular traits, climatic extremes would turn
living creatures over to their Maker at unprecedented rates."[1]

Water has been held sacred by virtually all the world's religious and
spiritual traditions. As Chinese sage Lao-Tzu long ago observed, "The

11

sage's transformation of the world arises from solving the problem of water. If water is united, the human heart will be corrected. If water is pure and clean, the heart of the people will readily be unified and desirous of cleanliness. Even when the citizenry's heart is changed, their conduct will not be depraved. So the sage's government does not consist of talking to people and persuading them, family by family. The pivot is water."

Our very life originates in the water of the womb. Yet we have treated it as hardly more than an industrial "resource" and noxious dumping ground. We are learning that we cannot live without this precious substance, but it seems that perhaps we have had to defile this sacred essence of life in order to learn our intimate dependence on it, and to undertake the restoration of the world's waters.

THE PROBLEM

We are witnessing the disastrous effects of toxic agricultural and industrial practices, uncontrolled development, urban growth, "heroic" engineering projects that dam or divert nearly all rivers, and rapacious overfishing. Sewage and toxic chemicals poison rivers, lakes, coastal areas, and even the high seas, introducing virulent toxins into the global food chain. About 80 percent of diseases are transmitted by water. Aquifers that provide drinking water are falling, and wetlands are disappearing. Irrigated lands are declining in productivity and being lost to salinization. Fisheries are collapsing. Conflicts among nations and groups over access to shared rivers or fisheries are escalating, creating serious prospects of water wars.

Finally, global warming is a potential source of great instability if marine currents, rainfall, coastal flooding, and weather undergo as significant a shift in pattern as anticipated by many scientists. The nation of Holland is examining the feasibility of constructing a huge offshore dike to offset flooding in case global warming causes an expected rise in ocean levels! The entire state of Florida is near sea level, and even a small rise in the oceans could deluge it and innumerable coastal areas worldwide.

Global water use has quadrupled since 1940, but it is very unevenly distributed. Average per capita U.S. use (the highest in the world) is 7,200 liters a day; and in India it is 25 liters.[2] Around the world, water tables are falling: in Bangkok, they have fallen some eighty feet since 1958, and continue dropping some 12 feet a year; in Tamil Nadu, India, they have fallen some hundred feet in the last 20 years.[3] The giant Ogallala aquifer in the U.S. Central Plains, supplying 20 to 30 percent of all U.S. irrigation water, has been severely depleted, forcing major reductions in irrigated acreage, especially in Texas. China, with 22 percent of the world's population but only 8 percent of its freshwater supply, is now the world's fastest growing economy. But Beijing, Tianjin, and the North China Plain are all facing very serious water shortages.[4]

At least 26 countries are now considered "water scarce." Many of them are in the Middle East and Africa, with rapidly rising populations in already politically volatile regions, and they share with several neighbors river basins that are virtually their only source of water. In fact, 40 percent of the world's population lives in river basins shared by more than one nation.[5]

In the Third World, at least 1.5 billion people lack access to unpolluted drinking water. The resulting diseases kill millions annually (up to 25,000 a day). About 70 percent of India's water is polluted, and 41 out of 44 of China's large cities have polluted groundwater.[6]

Recent studies link chlorine, the very chemical used to disinfect water, to cancer. Sporadic crises erupt in urban water supplies—Milwaukee's infamous 1993 cryptosporidium outbreak, New York's scare that same summer, and high lead levels in many municipal water systems.

The Natural Resources Defense Council (NRDC), a nonprofit environmental watchdog, found 250,000 violations of the safe water drinking act in 1991–1992, when 43 percent of the nation's water systems had violations. A U.S. Public Interest Research Group (PIRG) study found 21 percent of the nation's largest industrial and public waste treatment facilities in chronic serious violation of their discharge permits, and another 19 percent with occasional violations. In 1993, some 31 states reported concentrations of toxic contaminants in freshwater fish tissues that exceeded public health standards.[7]

The damming and diverting of nearly every major watercourse on Earth has, along with pollution, had such a destructive effect on wetlands, lakes, watersheds, and all riparian habitats that freshwater aquatic species are the most threatened form of life on the planet. One-fifth of the world's freshwater fish are endangered or extinct.

13

Modern agriculture (see Chapter 5) is a water hog, the largest user and waster of global freshwater supplies. Poorly conceived irrigation projects and meat-centered agriculture are among the main culprits. Agriculture is also arguably the most polluting sector of the economy, dumping millions of tons of animal wastes, nitrates and phosphates from fertilizers, pesticides, herbicides, and fungicides into our water.

Although industrial "point source" pollution (traceable to one source) is now more strictly monitored in industrial nations, violations are still rampant, and the very products of industry pollute. Many synthetic plastics, detergents, fibers, solvents, and pesticides can be toxic and resist degradation. Phosphate contamination of water rose sevenfold from 1940 to 1970.[8] Carcinogenic PCBs are detectable in mother's milk throughout the world.

Even industries with reputations for modernity and "lab coat" cleanliness, such as the computer industry, turn out to be lethal polluters. California's Silicon Valley contains the largest concentration of hazardous waste cleanup sites in the U.S. (23 on the EPA Superfund list). Much of Santa Clara County's groundwater is contaminated with trichloroethylene (TCE) and other toxic compounds used by the computer industry. Miscarriages and birth defects among computer industry workers and "cancer clusters" in Silicon Valley neighborhoods have all been reported.[9]

Many people became aware of ocean pollution in recent years when high bacteria levels caused by overloaded sewage treatment plants, raw sewage discharges, polluted storm water runoff, faulty septic systems, and the dumping of boating wastes forced numerous local beaches to close temporarily. In the United States alone there were 484 beach closings in 1988, climbing steadily to 2,619 in 1992.[10] The coastal portions of oceans are the most threatened, and they are precisely the richest, most biodiverse regions of the sea. Around 90 percent of the world's fish catch is taken in the one-third of the oceans closest to coasts.

Is it any surprise that coral reefs throughout the world are in serious decline? Or that red, green, and brown algae toxic tides (phytoplankton blooms) are occurring far more frequently, even in waters where they have never appeared before, resulting in tragedies such as the poisoning of 200 Guatemalans (including 26 fatalities) and the 1987–1988 deaths of up to 2,500 dolphins on the U.S. East Coast?[11] Or that polluted seafood in Peru caused a cholera epidemic that killed 3,000 people in 1991?[12]

14

All these signals point to a beleaguered water supply. The emerging consciousness of our dependence on the liquid biology of Earth is forcing a literal "sea change" in our attitudes toward water. Although the bioneers show that the solutions to these problems are within our reach, the very first step is to acknowledge the scope of our mismanagement of Earth's waters.

SOLUTIONS

We are learning to treat water as the precious substance it is. We are recognizing the immediate need to stop our massive dumping of toxic chemicals and sewage into the Cleopatra's bathwater of a finite, closed-loop system. Learning to stop "spitting in the soup," as President Lyndon Johnson once said, obviates the need for cleanup in the first place. Like stopping smoking cigarettes, the mere act of pollution prevention will cause a great surge in positive ecological health and save huge amounts of money.

Pricing water to reflect its true value encourages better use, a trend now emerging in European industry (see Chapter 7). Wasteful agricultural irrigation is in the process of radical reform around the world. Proven methods exist that could save 30 to 90 percent of agricultural water use. Reducing a water-consumptive, meat-centered diet would also drastically reduce water use and pollution. Many cities are starting to mandate water-saving household appliances, such as low-flush toilets. Numerous southwestern communities are also encouraging xeriscaping (dry landscaping), and discouraging water-consumptive lawns and golf courses.

We can restore aquatic biodiversity by stopping coastal overdevelopment and curtailing massive dam and irrigation projects with smaller scale technologies such as microdams, low-tech solar pumps, and shallow wells. Wherever these simple technologies are applied, almost instantaneous results are visible. The precipitous drop in fish catches from overfishing worldwide is starting to force a global transformation toward ecological ocean management. After threatening their own source of livelihood by overfishing, fishermen in New England are being compelled to adopt harvesting standards within ecological limits.

15

WHAT YOU CAN DO

As an individual, you can use water-saving devices for home and garden that will greatly reduce your demand on our finite water supply. Home systems are becoming available like the ones described in this chapter for natural water treatment. You can join a local watershed or environmental group to learn how water defines your region. Gaining the direct experience of knowing how water shapes the ecology of your place will change the way you experience the world.

You can bring your knowledge of natural treatment systems to your municipality or local industries. They will save money while cleaning up the local environment by adopting such methods. (See the Resource Section for more information.)

As John Todd's work illustrates, and as Jennifer Greene later describes so eloquently in Chapter 8, world cultures have long associated water with human emotions and with the sacred oneness of all life. Water, the great dissolver, teaches us that we cannot finally separate ourselves from the environment.

Ultimately, as the bioneer visionaries interviewed in this chapter reveal, by learning from natural systems and emulating them cleverly, we can start to restore Earth's degraded waters. Both John Todd and Donald Hammer, whom you will meet in this chapter, have looked deeply into the heart of nature to discover the nature of water. They have emerged with practical solutions so dramatic and so simply elegant that their work has begun to flow into the mainstream. It seems inevitable that these natural treatment systems will continue to spread worldwide and help cleanse our water more effectively, cheaply, and aesthetically.

John Todd's Alternative Greenhouse Effect

John Todd has developed advanced systems of wastewater treatment based on simulated ecosystems and solar design. Imitating nature, he is able to treat even the most noxious substances and serve community needs on a wide range of wastes.

IN THE LUSH SOLAR GREENHOUSE OF THE NEW ALCHEMY INSTITUTE IN 1972, biologist John Todd was leaning with frustration over the large tank where he was growing fish. Why, he puzzled, couldn't he get it right? No matter how many times he adjusted the balance of the fish population and the solar exposure or temperature, the tank filled with fish poop, choking the life of the system. In a moment of abstract, almost mindless reflection, he grabbed a piece of Styrofoam remaining from a planter, plucked a mat of watercress from a nearby pond, stuck it through the Styrofoam and tossed the little raft in the tank. It bobbed and tacked and finally settled, floating over the swirling motions of the surprised denizens of the water.

When Todd returned the next day to empty the tank of fish wastes, he found the water clarified. He quickly comprehended that he had inadvertently linked with a natural cycle of profound significance. The watercress did not treat the fish poop as waste but as food. The plant was quickly absorbing the rich nutrients from the water to create green cells for its living body. The experience would later inform Todd's work to help him create "living machines" capable of a wide variety of human-directed tasks, including transforming wastewater into purified water.

NEW ALCHEMY: TURNING WASTE INTO GOLD

Todd, who cofounded the nonprofit New Alchemy Institute with his wife Nancy Jack Todd and inventor William McLarney, was already a pioneer

17

in the alchemy of taking the gold of sunlight and transmuting it into energy, food, and shelter. The institute became famous for its early forays into advanced solar design, intensive fish farming, organic gardening, and sustainable food production. But Todd, a Canadian limnologist—a freshwater specialist, a pond man—could never stray far from his first love of water. Spread over the shallow sand spit of Cape Cod, the breezy Massachusetts landscape is dappled with ponds yearning to join with the roiling Atlantic Ocean. During years exploring the ponds and designing ecological solar- and wind-powered seagoing boats, Todd found himself losing a number of close friends to cancer. Surrounded with death, he took another look at the toxic holes in the web of life through which his youthful friends seemed to be falling. "That led me to investigate the waste treatment industry," he recalls in a Canadian accent clipped like a tight sail.

It is 1990, and the high-tech solar greenhouse in Providence, Rhode Island, is anomalous next to the Dickensian red brick backdrop of this classic nineteenth-century New England industrial city. Directly adjacent is a large conventional sewage treatment facility and outdoor lagoon bordering the Narraghansett River, which feeds the once fertile Narraghansett Bay. Known as the costume jewelry capital of the world, Providence has the dubious distinction of pouring pounds of heavy metals each day into its waters, along with ample human biological wastes. These metals are enough to decommission the conventional chemical-mechanical treatment plant with unnerving and costly frequency.

The striking greenhouse flies a bright whole-Earth flag in defiance of this archaic setting. Inside the greenhouse, a cloth mural sporting a graceful turtle ripples softly over the bursting plant life growing out of water in tanks. Near the back of the building, a battery of tall clear cylinders is filled with dark water, the sewage from 150 households, which drains into another line of these odd translucent tubes, whose sides are now thriving with rich green algae. The tubes empty into a complex set of smaller tanks in which a startling diversity of plants is growing. The scents drift on the thick air, overwhelming the senses in a hanging garden of Babylon in water pots, the brazen red of rhododendron spilling lustily onto the cement floor.

The water then channels into a set of marsh grasses, reeds, and rushes. On one of the barrels at the bottom of the greenhouse, Todd grins and turns a small tap, drizzling clear water proudly into a cup. "Like a sip?" he queries impishly. "It is drinking water quality, but it does still need to be sterilized just for safety. We use ultraviolet light for that."

John Todd has indeed taken some of the blacker water around and turned it back to gold. The Providence project, which was funded by several foundations, has demonstrated the ability of Todd's "living machines" to treat water ecologically by mimicking natural ecosystems.

Todd, whose red hair is now turning gray around the temples, peers out from deep-set blue eyes, making his large head appear almost volcanic, a force of nature brimming with primal creative magma. "For twenty-five years now I have been exploring the natural world," he says with gentle intensity, "attempting to understand how nature acts as a designer, how it performs as an architect. What are its stories — the many thousands of little natural histories that occur around the planet? Is there pattern or meaning? Ecology is really the superb science of relationships. What do those relationships mean, and how do they work? At the core is evolution. Have there been things going on for the last four billion years on this planet that we need to know about? This journey is really nothing more than a quest for a set of instructions that will allow us to behave in this world. It is the source of a body of knowledge that is comprehensive, integrated, and beautifully woven together to create the foundation for what I call the Age of Ecology."

LIVING MACHINES

Almost whispering, Todd stalks the mystery of the natural world. "I've learned that it's possible to clean up some of the horrors that face us on this planet, to destroy some of the most toxic compounds, to sequester others, to take some of the most ravaged parts of the planet and render them whole again. The means to accomplish this are ecological technologies, which I call 'living machines.' They wed human inventiveness with the workings of nature and the wisdom of the wild. In a sense, the human behaves like a farmer."

First, says Todd, a living machine is alive. "Its parts are hundreds, even thousands, of species of living things, of microorganisms, plants, and animals, including the higher animals. It is powered by the wind and the sun." Solar energy is central to the system, and Todd's patented clear cylinders act as a large solar collector, a photosynthetic lens soaking in the sun's power.

"Like an inanimate machine," he continues patiently, "a living machine does work, but it does so with the intelligence of three and a half billion years of experimentation by life forms. A living machine is a consciously directed ecosystem aimed toward alleviating a human-caused problem. It is very unusual in that the human being in its partnership with the living machine is the junior partner, because the intelligence in the system is contained within the organisms themselves. They know how to self-design, self-replicate, and self-repair. It is possible to create a living machine that will last for hundreds or thousands of years, and to do so in a manner that need not exploit the natural or wild world which sustains all of us."

Indeed, Todd points with admiration to the green film on the upper tanks, microscopic plants that act in synergy with bacteria. "They can do things in combination with bacteria that bacteria can't do alone. The symbiotic association between algae and bacteria and other organisms produces a dynamic ecology. The complex microbial communities are the foundation of living machines." He notes that when celebrated microbiologist Lynn Margulis inspected his tanks, she discovered at least half a dozen microbial communities pressed tightly together operating at different wave lengths. She had never seen anything quite like it before, except perhaps in the salt flats of Baja California, known for its outstanding biological diversity. Todd adds that the rocks in the tanks also serve a geologic purpose. Mineral diversity is a key food source for the microbes and plants.

Holding his face close to the transparent edge of the tube, he praises one of his "workhorses," a snail. "Snails do so many interesting things. One thing they do is eat sludge. Sludge is very difficult to deal with conventionally. And when the snails die, their shells begin to degrade and provide the energy for the essential bacteria. Snails are considered a nuisance if they get into conventional waste treatment plants. But from our point of view, they're one of the keys."

Moving to the middle component of the system, Todd looks approvingly over the copycat marsh he has lovingly modeled. "It runs as a tidal marsh, being oxygen-filled for half the day and oxygen-deprived the other half. The pulsing and the steep gradients provide a dynamic quality much the same as the edge of the ocean. Pulse exchanges are things like day and night or heat and cold. Steep gradients are sharp points where the realms of oxygen and nonoxygen intersect, or where light and shadow are carefully interspersed. We also introduce organisms at different times of year so the intelligence of seasons can be put into the system itself."

Gliding to the next ecological module, the "fish room," Todd notes the wide variety of flowering plants. "The goal of our work in this system is to find plants which have several functions: to destroy pathogens, sequester metals, remove nutrients, or provide the habitats for microbes and fish." He plucks a piece of fresh mint, pricking its aroma into the thick botanical air. Mints are known for their ability to destroy microorganisms that are harmful to humans, including streptococcus. The marsh plants themselves produce natural antibiotics so that the water does not need sterilization at the end.

THE ECOLOGICAL PLAY

Apart from providing visual beauty, the many flowers also have another purpose. They act as a "phosphorous sink," meaning they accumulate the otherwise difficult-to-remove chemical from the water. Last, Todd points out with delight, come his beloved fish. "The organic material that passes through the fish intestines is exposed to intestinal organisms, which are marvelous at transforming what is a waste into flesh and bone. That funny little hole sticking to the side of the tank is actually the mouth of an Amazonian armored catfish, one of my favorite actors in the ecological play. If any sludge gets into the end of the system, it translates it into flesh and bone. It works around the clock and prefers nighttime duty. That tank was as clean the day we broke it down as the day we started it, despite all the nasty stuff that had come through."

Biological diversity is essential to the system. "One needs to have everything from the ancient creatures, such as the early anaerobic phototropic bacteria, to the vertebrates for these systems to run for any length of time. It's alive. It's created by a human, but it's not human." Moreover, this natural waste treatment plant is effectively a farm capable of growing economic plants, such as cut flowers for decoration or saplings for reforestation. The system also raises Golden Shiners, a baitfish with a convenient appetite for phosphorous.

Todd observes that the system is "built exactly the way you and I are. It is made up of cellular units that are in many respects autonomous and at the same time interdependent with the other systems around it. The systems are interconnected in order to exchange genetic information among themselves." Surveying the living machine, Todd sums it up bluntly. "It's beautiful, and it doesn't stink."

As the Earth flag flutters over John Todd's greenhouse effect, the conventional treatment plant next door, which meets society's discharge standards and is well run, nevertheless pours out toxins and sludge into the bay and is unsatisfactory for the surrounding aquatic life. Meanwhile, the living machine functioning in the greenhouse is exceeding secondary discharge standards under the watchful eyes of the totemic turtle rippling on the guardian flag over this water garden.

"WHICH OF YOU LIKES THIS MESS?"

Todd proved the viability of living machines with sewage, which constitutes a major environmental and public health issue in cities across the country and the world, where discharge overflows and impure water seep into groundwater, rivers, and oceans. Conventional treatment plants also add the chemical chlorine, which data suggest can cause cancer, to kill any viruses or pathogens. Todd's natural systems obviate the need for the noxious chemical.

The activist biologist then moved on to a tougher challenge. Septage, the rank stuff which honey wagons (trucks) drain out of septic tanks in nonsewered areas, is ten times as concentrated as sewage; it also constitutes a serious environment and health threat. Many a septic truck disappears down the road to a place few people ever contemplate, and discharges on a moonless night over the hill. Increasingly, municipal sewage plants are rejecting the loads because they clog the digestive plumbing of the plant and overload the system.

Cape Cod, like many rural and semirural communities, is a nonsewered place. "Cape Cod is a freshwater lens," Todd explains of his home turf. "All of our water comes from rain. Everything we do on the surface of the land eventually ends up as our cells, especially in a place like this, which is just a sand spit with little soil. We have one of the highest cancer rates in the country because of what we do on the land. You see these huge pits where all the septage from homes, small businesses, and recreational vehicles is dumped. When I saw one of these pits for the first time, all I could see was the faces of unborn children in it. When we analyzed it, it had four-teen of the EPA's 15 top priority pollutants in very high concentrations. On the surface fats and greases from restaurants were floating. The pit wasn't lined, and it only takes a matter of days for some of the material to soak down into the water table. This is the story of almost every town."

The cost of conventional treatment plants is high, at $25 million, well beyond the budgets of most smaller communities. Todd designed an inno-vative system of 21 sequential clear cylinders, 21 aquaria on a hillside con-nected to form a gravity-fed river. He then acted as a biological sleuth, plumbing many different local lakes, ponds, pools, and pig wallows in search of organisms. "We said, 'We don't know which of you organisms like this mess, but there may be some of you, and we certainly hope so.'"

The system again began with the ancient organisms, bacteria, and moved through to the higher plants, vertebrates, and fish. A local stream plant, the water star wart, thrived and climbed up and down the side of the tank. The bulrush family again proved itself effective at breaking down organic carcinogens. Out the other end of the 21 systems came clean water with low bacterial contaminant counts. It met drinking water standards for metals. Of the 14 priority pollutants, 13 were 100 percent removed, and the fourteenth fell to a very low level.

"It was then that the political leaders and the Board of Health of Har-wich and the citizens decided to go ahead against the regulatory commu-nity and wastewater establishment to have a commercial living machine to treat their waste. We built a small commercial facility designed to treat fifty percent of the waste on an experimental basis for the state of Massa-chusetts. The early atmosphere was so hostile and negative that it literally took two and a half million dollars of water chemistry and engineering to convince the people that this was a valid approach. Some folks were terrified by the thoughts of flowers, snails, and fish in a system. It was not in the engineering books."

Investors formed a company called Ecological Engineering Associates to commercialize the technology, but the enterprise has had to struggle continuously against the recalcitrant engineering industry, which builds and profits from large, expensive chemical-mechanical treatment systems. Four years after the original experiment, the technology was finally accepted as effective by the state.

Todd expanded into other ventures, designing a living machine for Ben and Jerry's ice cream factory in environmentally stringent Vermont to treat the company's fatty milk wastes. He undertook a similar task for the Body Shop, the British personal-care products company whose wastes were also highly problematic, filled with fats and oils. A further sewage treatment facility got underway with support from the EPA and the Massachusetts Foundation for Excellence. Located in Frederick, Maryland, the facility is near Washington, D.C., and Todd hopes its proximity will influence federal policy.

CLEANUP AT A SNAIL'S PACE

After Todd spoke at an environmental conference in Chattanooga, Tennessee, people in the audience confronted him with their own local crisis, asking whether he could treat "the really horrible stuff." What they were talking about were 42 separate abandoned hazardous waste sites, including a major Superfund priority site along the Chattanooga Creek in a poor African-American community. The mess included some of the worst industrial chemicals and the most difficult to degrade, including pesticides not soluble in water and complex molecules called polyaromatic hydrocarbons. The creek itself had over 8 feet of sediments made up of these noxious complex compounds. After a poignant visit to the creek, known as Chattanooga's Love Canal, where waste sites spilled over into children's playing fields, Todd told the local people whose lives were being shattered that he would give it a try.

The EPA was skeptical that Todd's approach would be either workable or cost-effective. There were 33 separate and extremely difficult materials in the sites. There was a hundred-year history of dumping various toxic compounds, coal-tar derivatives, creosotes, and pesticides along five and a half miles of river. When the project was rejected for funding by the EPA, Todd went to the city, the county, and a local foundation for money to do a pilot bench test. With a modest $70,000 he was able to start the experiments, and to run a parallel control test at another lab in Canada.

The officials accompanied Todd to the creek to get his samples. "They had taken me to perhaps the worst site in the whole creek. We pulled out a core that was just one inch thick, and it reeked so strongly that the TV crews leapt backwards because it was so foul smelling. The material was so toxic that the first living machine we built had to be outdoors and I could only stand upwind of it if I didn't breathe."

Then the biology kicked in. "At the point when the first cell began to recognize the chemicals, we were pulsing it through an aerobic/anaerobic cycle and got some things right. The first cell started communicating with the second cell, they started exchanging genetic information, and finally the stuff began to break down. Interestingly, the organisms that I got for the process came from the sea, from a salt marsh near where I live. I had this image of the ancient ocean beginning to clash head-on with the modern chemicals of you-know-who. I regulated the rate of exchange between cells by the rate at which snails laid eggs on the sides of the second cell. If they stopped laying eggs, I knew I was pushing too fast."

Moving at a snail's pace in a two-celled system achieved the desired results. Data from the project was promising. Some 21 of the 33 compounds were removed either 100 percent or at very high rates. These included some of the most toxic substances. Three of the compounds were barely removed. Nine others actually increased in concentrations, but they were all breakdown products of the higher, more complex compounds, and in most cases were less toxic. The parallel Canadian test registered similar results. "It moved me from an environment of despair to an environment of possibility," Todd concludes.

The experiment was intriguing enough for the EPA to request a comparative analysis of living machines with other technologies, but still no funds were forthcoming. Todd says the living machines would cost about $1 million for each half mile of creek, or perhaps $12 million altogether, compared against the conventional approach of dredging and incinerating for $80 million. In response, the business community of Chattanooga pledged to put up 50 percent of the money if the EPA would match it. The situation has remained at a stalemate. But, Todd reports, the whole school system of the city, from kindergarten through high school, is using living machines as teaching tools to impart ecological literacy to the next generations.

LIVING LIGHTLY

Meanwhile, Todd continues to generate a lava flow of new ideas and applications. Among the most innovative is a "pond restorer," a floating island that serves to purify lakes and ponds. Demonstrated on Flax Pond in Harwich, Massachusetts, the odd contraption was inspired by a freshwater bryzone, a spongelike animal with a structure similar to coral reefs. The biological pattern of the sponge is designed by nature to provide multiple surfaces covered with bacteria over which water washes and cleanses itself by providing nutrients to the microbes living in the sponge. Todd's system, which looks like an ancient Spanish galleon, upwells water from the pond using two electrical-generating windmills, and passes it over and through a bladderlike mechanism based on natural design.

Flax Pond had been closed for years from recreation because of priority pollutants and coliform bacteria from human wastes. With funding from the town of Harwich, the EPA, and the Massachusetts Foundation for Excellence, the living machine has now begun to restore the water, and the pond is once again open for fishing and swimming. The system is now being enhanced to treat one-quarter million gallons a day, and Todd is in discussions with the city of Toronto, Canada, to restore the city's once beautiful beaches for swimming, which hasn't been possible for 20 years.

"Nature can design at four orders of magnitude or ten thousand or more efficiency than human engineers," Todd notes triumphantly. "That's the theoretical range of improvement we can achieve if we listen to the internal architectural teaching of the natural world. It is now possible," he adds boldly, "for humans to live lightly on the planet using only one-tenth of what contemporary society uses. Such an improvement would represent three or four orders of magnitude. It may be possible in the next century for us to reduce our impact on the natural world by ninety percent. That would mean we could give back to the wild the liberation of nature from us. We could return ninety percent of the planet and still have high culture. We need to give that legacy back."

THE BARD OF BIOLOGY

Intrepid John Todd continues his quest to restore the world's waters, a bard of biology with a vision of harmony. He is applying the work now to treating wastewater from boats, a waste so toxic that it's deadly to most life forms. He groans at the 77,000 acres of shellfish beds lost to New Englanders from such trivial yet catastrophic wastes. Because boats operate mainly in salt water, he had to identify a community of organisms that like brackish water. Within a matter of months, he was producing clean water from toxic wakes.

Todd is now turning amphibious. The wastewater facility in Providence, Rhode Island, ended its successful demonstration, but the city chose not to implement the living machine. So Todd is converting it to produce food in water gardens, which he envisions as the urban fresh market of the future, growing organic produce and of course fish.

Todd is convinced that living machines can not only treat wastes and grow food without harming the environment, but also generate clean fuels to heat or cool buildings. "If you were to redesign cities and towns," he envisions, "so that buildings treated their own wastes and heated themselves, so that streets were farms as well as meeting places, so that some streets were dug up and made into ponds, and where living machines on the edges of ponds kept them pure without the use of chemicals, imagine cities of the future which were themselves unplugged.

"The twentieth century undoubtedly will be remembered as the time when technology was out of balance. It is time to take a whole new set of technologies, with an attitude and a spiritual approach to transform them into the whole fabric of our communities as living ecologies. I believe we're on a cusp. We will either continue business as usual, or rethink our culture and find our instructions and direction in the life and beauty and deep integrity of nature."

"We're building intentional ecosystems," he elaborates with fascination. "Nature is a symphony which is ultimately unknowable. You come out with a world that's much more tentative and relationship-based, much more mysterious, powerful, and dynamic. Nature orchestrates. It's symphonic. There are single notes, but there are no soloists."

AS IF IT WERE THE LAST DAY ON EARTH

Todd's humility is genuine, and he shares a Tibetan Buddhist practice he adopted some years ago. "When my mother got very ill, and it was just a question of when rather than whether, I began to live each day as if it were my last day on Earth. That was the most revolutionary step I could take. Let's say it's your last day on Earth. First of all, you don't want to spend it with people you don't like or with whom you are in deep conflict unless resolution of that conflict that day can make a difference. Second, you don't want to work on the trivial. So if you're really feeling a little lazy and you've got three or four phone calls from people that really don't matter in your scheme of things, you don't touch it.

"You're doing a kind of closure with all your close relationships, where there are going to be no harsh words. It's your last chance with whomever you work with and live with and love. Although you may sometimes forget about it and slip, slowly, over time, you learn to live each day as the last, and you change. You start to become more courageous because it's your last day. You're willing to make mistakes."

As a biologist, John Todd sees that the world's waters are very sick, and he is acting as though it were their last days. For that very reason, they may well recover in our lifetime.

Donald Hammer's Shallow Ecology

Dr. Donald Hammer has pioneered the use of Natural Treatment Systems for wastewater treatment around the world. These constructed wetlands are more effective and cheaper than conventional treatment for purifying wastewater from almost all sources. Their widespread application will dramatically enhance water quality while conserving water.

THE GENERIC MOTEL ROOM IS LIKE A THOUSAND OTHERS ANYWHERE ACROSS the United States, bristling with plastic fibers and synthetic scents. The TV is tuned to the weather channel, which Don Hammer watches tensely. He loosens with relief at the news that the rainy weather will let up soon. It has been pelting for days here in Alabama, and the ecologist is concerned about how the constructed wetlands he recently installed will handle this much liquid input before the plants have had a chance to mature.

Indeed, if you hear a sucking sound in the South, it could well be Hammer's boots slogging through one of hundreds of such wetlands whose construction he has overseen. A constructed wetlands is different from a natural or restored wetlands by virtue of its specific application for the treatment of wastewater. While natural wetlands, also known as marshes or swamps, serve many functions, including wildlife habitat and flood control, they also serve as nature's living filters to purify water.

"We have always dumped our wastes into water," says Hammer in a gravelly voice as he surveys the verdant marsh brimming with cattails, rushes, and reeds. The human-made marsh sits below the standard package-treatment plant, a mini-sewage system at the Bear Creek High School in northwest Alabama. The conventional system had problems with noxious discharges above ground and into the water table. The installed marsh system has now acted as a "polishing" step to remove or transform the pollutants. "The biology teacher now uses it for his lab exercises," Hammer says with wry satisfaction. "It works just as well Monday through Friday as it does on Saturday or Sunday without students. It works just as well during July and August, when school is not in session, as it does in January, when

29

school is in session with a substantial loading difference. Even with 800 kids in the school, the system is amenable to widely fluctuating loading rates. Biological systems are much more flexible."

NATURE'S BIOTECHNOLOGY

Hammer, a restoration ecologist who once specialized in reestablishing nesting populations of raptors such as ospreys, hawks, and eagles, has since emerged as probably the world's foremost authority on the use of constructed wetlands for wastewater treatment. He literally wrote the book on it, a three-pound tome that serves as the chapter and verse on state-of-the-art knowledge of this biological technology.[13] In fact, constructed wetlands for bioremediation may well be the most authentic and exciting biotechnology of the emerging Age of Biology.

Ironically, a curious group of bioneers like Hammer more or less synchronously stumbled across the bioremediating properties of constructed wetlands. Several of them were motivated by concern over the dumping of highly noxious wastes from coal strip-mining operations into pristine natural wetlands. The red water, thick with iron and other heavy metals, was flooding unchecked into the rich marshes of the southeastern United States. When these eco-detectives staked out an area and commenced measuring the damage they expected to find, they discovered, to their surprise, that the water came out clean at the other end of the marsh.

The biologists figured out that the wetlands were purifying the water. The reason, they realized, is that in nature there is no waste. Everything is something's "lunch" of food or energy, and one organism's poison is another's food. So bioremediation using natural treatment systems becomes a game of mixing and matching ecologies of organisms to digest a particular substance, transforming "waste" into resource. Indeed, the whole notion of "wastewater" itself belies the wrongheaded antibiological attitude people have held toward this invaluable resource. We are literally wasting our water and our nutrient resources along with it.

The bioremediation technology of constructed wetlands is biomorphic, that is, it imitates and mimics natural processes to achieve desired biological results. Simulating natural wetlands ecosystems has proven the surest way to replicate these cleansing processes. Unfortunately, only 40 percent of U.S. natural wetlands still remain, as human encroachment has steadily diminished this vital natural ecology. At the same time, human population has increased radically, and natural wetlands have suffered and retreated from spreading human development, including houses, agriculture, and the draining of marshy areas.

Hammer has spent years empirically testing the designs of these systems, and he takes a hard-nosed scientific attitude toward their effectiveness. The ones that work best are called "surface flow," and they function by having the water flow across a shallow bed of soil. Alternatively, systems that use a subsurface flow and gravel have not worked adequately, and Hammer grows impatient with those who promote such models.

Hammer, who has traveled the world installing constructed wetlands in virtually every climate and ecology in the world, got his start through the Tennessee Valley Authority (TVA), the huge federal utility which hired him in 1972 to conduct a large Canada goose restoration program. While leading a wildlife class in New England, he visited the Brookhaven national lab where Max Small was experimenting with constructed wetlands. Part of a small circle of such bioneers, Small evolved what is known as the "marsh-pond-meadow" system, a simulation of the naturally occurring ecology whose detoxifying function observers have compared to the "kidneys" of nature.

Hammer became fascinated by the system, and in 1979 obtained board authorization from TVA for a demonstration. Funding was marginal, however, an orphan in the giant steel-and-concrete pantheon of TVA engineering icons. But Hammer was undeterred and managed to squeeze through a few small projects, including municipal sewage pilots in 1983. Then in 1984, the inevitable occurred.

BEAVER SHOWS THE WAY

Because of the TVA's involvement in utility power generation and its location in the coal strip-mining country of the Southeast, the power folks asked Hammer whether these wetlands worked for acid drainage from mining. Hammer had seen such projects while camping. They were constructed by beavers, but he didn't mention that fact.

31

"Red water from acid mine drainage, or AMD, detrimentally impacts over eleven thousand miles of streams in Appalachia alone," Hammer cites matter-of-factly, stroking his Amish-style beard. "Coal companies currently spend over one million dollars a day in conventional AMD pollution abatement. Conventional treatment is typically caustic soda, which raises the pH and precipitates out the irons. But the maintenance is very high because the 'good ole boys' come up every Sunday afternoon and see if they can put a slug through the tank. You come up on Monday morning to find the caustic soda all over the road or the stream.

"When we noticed that natural wetlands including a beaver dam had red water coming in and clean water coming out, we thought, 'If a beaver can do that, we can do that.' But when a bureaucracy builds a beaver dam, it is different from when a beaver does. After the 'dozers got through, we went in and planted cattails and bulrush. Throughout the summer and fall, a pretty little marsh-pond complex changed a dead acid stream into almost trout-clean water, with a dramatic improvement in the iron, manganese, and solids."

The system incorporated a little riprap structure, a weir similar to a beaver dam, to help manage and manipulate the water level. "We learned this trick from the beaver," says Hammer approvingly, "and then the beaver moved in and started taking over the management of the AMD treatment system. I used to say that we couldn't convince the regulatory agencies to transfer the permit responsibility to Mr. and Mrs. Beaver, but we have gotten total releases on a couple of systems now."

There are well over 800 such constructed wetlands in Appalachia today treating acid mine drainage, and they are spreading rapidly around the world. TVA has installed close to 20 on its own operations.

Following Hammer's successful demonstrations for acid mine drainage and a municipal sewage system, TVA asked him to lead four teams examining those areas as well as industrial and agricultural systems. He expanded into three towns in western Kentucky to show the viability for treating municipal sewage. The towns of Benton, Hardin, and Pembroke all resolved their water-pollution problems quickly and inexpensively using natural treatment systems Hammer designed and installed.

One site had a special problem because of the severity of slopes in the little Appalachian mountain village. Whenever it rains, which happens often, the storm water runoff mixes with the shallow sewage and spreads dirty water around. The wetlands proved to be viable even on steeply terraced hillsides, and helped slow the rush of storm water while cleaning it at the same time. In fact, treating storm water runoff has proven to be one of the most important uses for the technology, and some 200 constructed wetlands have been installed for storm water treatment in Florida alone.

The technology also took hold abroad. The pioneering work of Kathe Seidel at the Max Planck Institute in Germany in the 1950s, which documented the removal of pollutants from wastewater by marsh plants, led one of her students to examine the effects of a natural wetlands in the village of Othfresen, Germany. The study resulted in the widespread popularization of the technology in Europe, where 200 such municipal and industrial systems are now functioning.

SAVING CASH AND FLOW

According to Hammer and to other field studies, the cost of installing a constructed wetlands is about one-third to one-half the cost of a conventional wastewater treatment system. The operating costs are only one-tenth to one-half as much, and the systems do not involve the use of costly and often dangerous chemicals. Rather than a highly trained engineer to operate the system, a gardener will do.

Cost is no small factor in the future of water quality. In the United States, most current conventional waste treatment systems were constructed in the early 1970s during a boom in federal Environmental Protection Agency (EPA) funding following the Clean Water Act of 1972 passed under Richard Nixon. Because the federal government largely paid outright for the systems, municipalities competed to obtain maximum funds to build expensive plants. These systems, however, have a life span of only about 20 years before extensive retooling is required. When Ronald Reagan put a reduction valve on federal funds in the late 1980s and changed the grants to loans, suddenly the overwhelming financial burden fell on struggling states and cities. For smaller municipalities, the costs were prohibitive, and they were the first to try biological systems out of economic necessity. For many countries in the world, especially so-called lesser developed nations, the costs of conventional systems are altogether un-thinkable, and even a relatively prosperous country such as Mexico has exactly one conventional plant. Consequently, constructed wetlands rep-resent the only viable option for many situations.

The spectrum of applications for this natural biotechnology is broad and impressive, and Hammer, who was appointed project director of the TVA Waste Technology Program in 1988, has worked with virtually all of them. Between Chattanooga and Knoxville, Tennessee, one of the largest pulp and paper mills in North America was discharging colored water like dark tea, filled with nasty pollutants including lignins and tannins, which are very hard to remove by conventional means. The experimental natural treatment system Hammer installed reduced pollutant counts from 3,000 to 100.

Another great unresolved problem is agricultural runoff, both from row crops excessively treated with nitrogen fertilizers and also from concen-trated animal wastes. In northern Maine, Hammer consulted with a for-mer student grappling with heavily fertilized potato fields adjacent to two deep-water ponds filled with landlocked salmon and trout in a valuable fishery. "The nutrient running off those fields into the lakes was destroy-ing the fishery through eutrophication [suffocation]," Hammer recalls. Di-recting the runoff through a wetlands complex removed 96 to 98 percent of the nitrogen and phosphate from the row-crop fields. "Very cheap," Hammer concludes. "Four or five thousand dollars, about fifteen to twenty thousand dollars for a watershed, very easy to build. Today it's a nice little marsh with a brood of black ducks on it."

Livestock waste is a principal threat to water quality as well. Whereas once the meadow muffins of cows were spread out over wide expanses, modern production methods concentrate the animals into small areas, where the runoff goes into thick lagoons. The lagoons have seldom been maintained properly by farmers, and discharge frequently into nearby streams. "Livestock waste introduces eleven to twelve times the nutrient loading into our reservoirs in the Tennessee Valley that human waste does," Hammer notes somberly. He examined a poultry operation with houses for 100,000 laying hens. The ammonia-nitrogen level was almost phytotoxic, enough to kill plants.

In 1988 Hammer worked with Auburn University to lay out an ex-perimental system for animal wastes, mainly hogs. The station set up a number of cells with wetlands plants. As long as the loading rates were not extreme, the wetlands did the job, sending clean water out the other end.

GROWING QUALITY OF WILDLIFE

In the frozen plains of Mandan, North Dakota, the Amoco oil refinery was having big problems with its discharges. Along with the complex mixed chemicals from the large oil-refining operation, the plant also produced human sewage and had trouble with storm water runoff, which spread the noxious brew across the landscape. Hammer was eminently pleased to hear that the constructed wetlands did the job of producing clean water, but he was perplexed to learn that the company was spending $60,000 a year to maintain it. He inquired about the high cost and learned that plant manager Don Litchfield spent the money hiring contractors to build goose-nesting platforms and plant sorghum and other grains to feed the flocks of birds and wildlife that now frequented the lush feeding grounds. On the lower end of the wetlands, he was also growing trout. "He was not paying anything for wastewater treatment." Hammer chuckles.

Another big problem across the country is failing septic systems on private property which leach out on the surface when they overflow. Hammer has applied constructed wetlands to these, using not only cattails and bulrushes, but also ornamental flowers. "You get a combination wastewater treatment system and a flower garden to beautify the homeowner's backyard." He smiles. Increasing numbers of homeowners are employing such polishing systems, providing advanced water quality treatment, beauty, and wildlife habitat to their own backyards.

OUR FRIEND THE MICROBE

Natural treatment systems using constructed wetlands have now been successfully demonstrated on the wastewater from textile mills, fish-rearing ponds, photochemical laboratories, seafood processing plants, compost leachate, landfill leachate, and sugar beet processing plants. It appears likely that they will work on almost all wastewater streams. They can also be applied effectively as the polishing or tertiary step after conventional treatment in municipal sewage systems, which have great difficulty in achieving this final 5 percent of cleanup without exorbitant cost and intensive chemical usage.

How do the systems work? "All we are doing with constructed wetlands is creating the habitat for microbial populations," Hammer says. "Plants do not remove pollutants. Wetlands do not remove pollutants. Microbes transform the pollutants. They are ubiquitous. Give them an energy source, and they'll be there. Bacteria, fungi, protozoa, and some algae provide the treatment process." He emphasizes that we know little about the microbial action that performs these vital digestive functions.

Hammer also points out that there are simple physical-chemical processes involved. If the pH, the balance between an acid or alkaline state, is high enough, iron will precipitate out, as will manganese and other metals.

What is the function of the plants? "The plants' main purpose is to die," says bioexistentialist Hammer. "They then provide a layer or substrate of detritus that becomes a habitat for the microbes." The other purpose of the plants is to provide oxygen to the underworld. Since wetlands plants are built like drinking straws, hollow in the middle, they are able to transport oxygen to the plants' roots and create an aerobic region around the roots, which combines with an anaerobic zone to support the kinds of transformations the bacteria accomplish. Little work has been done on using a greater variety of plants, and Hammer believes that the best candidates are native plants that are well adapted to a given region. Whatever will thrive there provides the best habitat for the bacteria.

How long will these systems work? Hammer recounts his experience in the high mountains near Yellowstone Park. At 11,000 feet, he saw a small red water seep that created a sedge meadow, a small wetlands that had clean water coming out of it. He points out that this natural process has been going on for millions of years. It works in the winter. "Wetlands have always treated wastewater. They occur in topographic low spots because water runs downhill. It isn't clean water, even if there are no people there. Wetlands systems have adapted over millions of years to process, transform, and thrive on these inputs of nutrients and energy. This is one of the reasons they are one of the most productive systems known. They

36

outdo a prime Iowa cornfield. They are naturally adapted to processing substances coming in through wastewater. All we have to do is learn to do it right."

The main obstacle Hammer has faced in the wider adoption of constructed wetlands is the industry of conventional wastewater treatment. "The engineers weren't trained in this technology in school," he notes evenly. "They are unfamiliar with it and resist it. More importantly, engineers and engineering firms survive on the basis of making twenty percent of project costs. They're not interested in twenty percent of one hundred thousand dollars if they can convince a community to build a four million dollar plant. I don't see that that's going to change. There just isn't enough profit in constructed wetlands."

Constructed wetlands do have limitations. They are not efficient in dense urban areas where land is at a premium. They require sufficient acreage to function properly, not an option in Manhattan.

YES, IN MY BACKYARD

Nevertheless, the use of constructed wetlands has spread rapidly, propelled by popular demand, and Hammer has been central to their burgeoning grassroots use. There are now close to 300 such systems in use for treating municipal sewage, over 1,000 private household systems, 800 acid mine drainage models, 30 for agriculture and 45 for industry.

According to Woody Reed, an engineer who conducted the EPA survey on the natural technology, the trend runs counter to the usual rejection by communities of waste treatment facilities in their neighborhoods. People like the cost savings as well as the creation of a "green zone" that enhances their landscape. Among the best known examples is the artificial marsh created by the city of Arcata, California, which not only has dramatically improved water quality as it flows from the town into the sensitive estuary of Humboldt Bay, but has created a large local park. Citizens often take a picnic lunch to the local waste-treatment wetlands where they go bird watching.

For Don Hammer, learning to use nature to heal nature has been an intensely personal journey. Growing up in the Prairie Pothole region of

North Dakota on a small grain and livestock farm, he looked out his bed-room window on the rich and diverse waterholes of the northern desert. In a region that received a scant 12 to 14 inches of annual precipitation, the potholes sprang to life with the coming of the rains. "I was fascinated by the explosive abundance of life in these systems from as early as I can remember. I spent much of my free time observing and hunting in and around the potholes. I was only interested in wildlife by the time I started university."

Attending the University of Utah, where he obtained a Ph.D. in ecol-ogy, Hammer got his first job trapping, ear-tagging and releasing mink for a movement study on a national wildlife study for the U.S. Fish and Wildlife Service in North Dakota. He studied turtles and their predators, and later taught wetlands ecology at the University of Maine. After joining TVA in 1972, he received numerous awards for his conservation efforts, including "Conservationist of the Year" from the Tennessee Conservation League.

As the welcome sight of cattails, reeds, and rushes spreads across the nation, Hammer is looking to the future. He left TVA in 1995 to pursue his work in the private sector, and continues to travel the world equipping people in this vital and simple technology. He has instigated projects in Thailand, India, and Mexico, where sewage treatment is nothing short of an emergency with open ditches that dump into rivers, oceans, and water tables. The community in Thailand was so excited at its new installation that the local government held a ceremony complete with a flaming mar-quee of fireworks spelling out a pyrotechnic "Constructed Wetlands." Ham-mer believes that the cost-effectiveness of the systems makes them an inevitable choice for many less developed countries around the world, and also presents an excellent business opportunity for farseeing environmen-tal entrepreneurs.

Hammer predicts a continuing proliferation of constructed wetlands because of environmental regulations from NAFTA and other interna-tional trade agreements. He points out that acceptance into the newly formed European Union requires improved wastewater treatment. "A pos-itive future to me means that I could drink water out of the tap anywhere in the world, and we would have substantially increased the amount of wetlands habitat and related wildlife."

Don Hammer has to get off the phone right now. His sister just called from the road, trapped behind a chemical spill on I-75, a few hours north of his Tennessee home, where she is coming for the holiday. Where will that spill drain? In whose drinking water will it end up? What if there were a constructed wetlands nearby to handle that nasty storm water runoff? Hammer turns back to the weather channel, hoping the rains will hold off for a spell.

CHAPTER TWO

The Web of Kinship

BIOLOGISTS GARY PAUL NABHAN AND STEPHEN BUCHMANN HAD BEEN visiting the managed hives of wild honeybees in the mountains outside Tucson, Arizona, for ten years. Each trek became more dispiriting. Nature's "most productive workers," who knew where every local wildflower was and when any one of these 1,500 flower species would bloom, were steadily losing the terrain on which they relied. The striated mounds of thousands of pollen pellets meticulously gathered and assembled by the bees reflected in miniature a landscape in radical decline.

Nabhan began to realize that a convergence of forces was conspiring to disrupt the pollination cycle itself, a biological process of fertilization of such keystone importance that it is difficult even to conceive of the consequences. At the same time that human development was steadily diminishing wild lands and native plants, the introduction of highly efficient commercial honeybees was also severely impacting the wild bee populations.

The commercial honeybees outperformed their wild relatives, leaving little pollen and nectar for them, and threatening their livelihood. Then an onslaught of pesticides, mites, and other hazards devastated the honeybee population and the U.S. bee industry. Finally, the arrival of Africanized honeybees added to the biological deconstruction. In one part of Arizona alone, the Africanized bees wiped out 85 percent of the remaining wild bees, which had spent untold millennia attuning to the intricate cycles of pollination and support of the fertility of the land.

The irony is that the pollination provided even by commercial honeybees is worth 50 or 60 times as much as their honey and other products such as bee pollen and wax combined. A survey of 60 agricultural crops grown in or imported into the United States found that seven crops worth about \$1.25 billion are pollinated mainly by wild insects. Nabhan points out that the pollination cycle is very poorly understood: We know only about one pollinator for any of 15 plants.

Field biologist Nabhan warns that the wild creatures and places hold vital essences critical to our survival and well-being. The farther that farmers' fields become separated from wild lands, the more their yields suffer. In other words, the knee bone's connected to the thighbone. The food "chain" is in truth a food web, vastly interconnected through horizontal integration. When only one plant becomes extinct, along with it will disappear the 20 to 40 animal and insect species that rely on it. Life is an elaborate tapestry, and biodiversity its very fabric. "One must begin," Nabhan urges us, "to see the world through the eyes of a bee or a butterfly, to smell out its fragrances as a moth would, to taste the mix of sugars as a hummingbird might taste them."[1]

Biological diversity is the actual stuff of life, encompassing life's myriad adaptations. Biodiversity constitutes the incredibly specific and idiosyncratic responses that life has adopted over 3 or 4 billion years of evolution. The gene pool is life's information bank, the repertoire it draws on in the wondrous continuous dance of creative evolution. It is mysterious and deep beyond our comprehension. We do not even know how many species there are in the world, or how they relate to one another, much less how they fit together in the vast tapestry we call life.

Many scientists today consider the loss of biological diversity to be the top global environmental threat. Biodiversity—the sum total of all the world's life forms, organisms, and genes—is nature's fail-safe mechanism against extinction. Once organisms and their genes are lost, they cannot be recovered. Extinction is indeed forever. Any smart banker will recommend a diversified portfolio to hedge against risk, a model that comes from ecology. Only lawyers and pests favor centralization, one biologist wryly remarked.

The one constant in nature is change, and in practical terms, biodiversity is the deck of options that represents our very ability to adapt to change. As we diminish biodiversity, we are no longer playing with a full deck.

THE PROBLEM

We are shredding the very fabric of life, and impoverishing creation. As human beings, large animals at the supposed top of the food "chain," we also depend on biodiversity. Yet we are radically limiting our own future range of options, jeopardizing our survival as a species and disposing of many others. Usually the principal motive is shortsighted economic gain.

Human beings have become 100 times more numerous than any other land animal of comparable size in the history of the planet. We appropriate an astonishing 40 percent of all the solar energy captured in land plants, sharply reducing the resources available to other species. Extinctions today are estimated at 1,000 times the natural rate, about 27,000 a year, leading to a massive species loss similar in scope to the five major such episodes in the last 500 million years. But unlike the previous huge die-offs caused by climate, geology, or stray meteors, this current crisis is the result of human activity. After each of these previous periods, it took life roughly 10 million years to reestablish a comparable level of diversity, albeit with a very different mix of species each time. Life itself will regenerate no matter what we do, but at a pace of 10 million years, time is not on our side.

The driving forces of this spate of extinctions are economic gain, "development," the degradation of wild habitats, the introduction of alien species into ecosystems, massive pollution, climate change, and the overharvesting of fish, trees, wild animals, plants, and other organisms. Sadly, we don't know even a fraction of the species on our planet. We know virtually nothing about the microorganisms in soil and their relationship to plant growth or ecology. Scientists do not celebrate when they discover a new species. They add it to the pile of organisms about which we seldom claim to know more than a name and classification, if that.

Habitat degradation is the current leading cause of biodiversity loss. The situation is bleakest in the most diverse and richest ecosystems—the tropics, especially rain forests where an estimated one-half to two-thirds of the world's species reside. Rain forests are already well over half destroyed and are disappearing at the rate of a football field per second. Tropical "dry" forests are even more endangered, and temperate and northern forests are not faring much better.

There are very tangible reasons we need a varied gene pool in agriculture and medicine. Wild species are the ancestors of all our food crops and continue to provide a vital reservoir of genetic traits from which we have consistently drawn to improve crop strains or to prevent or remedy disasters. When a pest or climatic shift jeopardizes a prevailing crop, a wild strain frequently has a genetic feature that makes it resistant to that problem, and it can be crossbred with the prevailing crop. Often, however, this trait is rare and found in few or only one strain. An obscure Andean potato rescued the Irish potato crop after the 1845 potato famine. A rare and "weak" rice species "discovered" in India in 1966 turned out to be the only one of 6,000 assayed to have genetic resistance to the devastating grassy stunt virus, and it is now grown all over Asia.[2]

The loss of biodiversity in agricultural crops is staggering and dangerous. Vice President Gore has called this genetic erosion "the most serious threat to our food supply." Since the beginning of the century, 75 percent of the genetic diversity of the world's food crops has been lost. Of 7,000 or so types of plants that have been used for food by humans, only 20 species now provide 90 percent of the world's food. Merely three of these – wheat, corn, and rice—account for 50 percent. Yet from 30,000 to 80,000 species of plants have some edible parts.

The world scientific community has issued a global red alert on biodiversity loss, and there is far greater awareness today within government and business of the magnitude of this true crisis. Many more citizens are also increasingly alerted, and positive actions are springing up in backyards as well as boardrooms to save the extraordinary diversity of life, and ourselves with it.

SOLUTIONS

To stanch biodiversity loss, our species is rapidly evolving a new maturity founded in a recognition of our place within the greater web of life. We are learning about our interdependence from nature's own superb technology and from the patiently acquired Earth wisdom of indigenous cultures which have coexisted successfully with life's myriad expressions for tens of thousands of years.

Habitat protection is of the first importance. Even from a purely utilitarian "bottom line" perspective, it is becoming clear that preserving richly biodiverse ecosystems and using them intelligently is far more economically sensible than replacing complex ecologies with monocultures such as cotton fields, shrimp farms, cattle ranches, or tree plantations. The sustainable extraction of Brazil nuts, palm hearts, tonka beans, and rubber by Brazilian tree tappers is proving to be more productive than clear-cutting forests. Some global financial interests are awakening to these realities (see Chapter 6).

One leading edge of habitat protection is the recognition of "biological corridors" through which wildlife move. Just setting aside islands of land, even sizable ones, as wilderness reserves is insufficient, since many creatures move around along natural pathways over very large areas. Many biologists and environmental groups are attempting to gain the ear of policy makers and industry with this knowledge.

Ecotourism and recreational use are also often far more profitable than the destructive short-term extraction of resources. In the United States, 95 million people enjoy outdoor recreation in wildlife areas annually, while another 70 million hunt and fish. Americans spend $37 billion a year on recreation in these areas. In Kenya, tourism brought $300 million in 1985. A male lion living for seven years there is estimated to be worth $515,000 to tourism, versus $1,000 for its skin. In Colorado today, tourism is more profitable than strip-mining, and who wants to ski down a slag heap?[3]

The reintroduction of diversity into the food system is starting to occur on a small but significant scale. Some progressive natural foods markets now offer as many as 44 types of heirloom beans and many "new" varieties of potatoes. Farmers are finding valuable crops in heirloom vegetables and fruits. The corporate domination of agriculture and control of seed stocks is being challenged worldwide with decentralized systems, an inspiring model described by Vandana Shiva in this chapter.

A recognition is spreading that seed banks, seed collections, and botanical gardens must be expanded and funded throughout the world, and local botanists trained. The study of our planet's species is being accelerated. The mapping of life forms, and the preservation of as many as possible, is gaining priority around the world. Companies large and small are working with local governments and peoples to prevent further loss in hopes of profiting from the living treasure of biodiversity.

Important strides are being made to acknowledge the profound contribution of indigenous peoples and Third World countries to preserving biodiversity with both compensation and cultural recognition. To achieve these ends, local people are being empowered to benefit from and protect their own healthy ecosystems. However, the patenting of life forms and plant species is a complex area of dispute, where strong arguments contend that such patenting must be reassessed altogether, and at times curtailed or disallowed if we hope to preserve biodiversity.

Economic incentives and disincentives are also emerging as effective tools. If the price of commodities can be adjusted to reflect more accurately their real worth and the impact of their extraction, markets can behave more rationally. "Green taxes" and user fees are one way to encourage this goal. (See Chapter 6 and Chapter 7.)

In all these areas, the struggle to restore biodiversity is being carried by grassroots organizations, tribal movements, human rights and environmental groups, botanists, scientists, authors, celebrities, and progressive business and international leaders. Some fight to protect endangered species and peoples, some set up models of habitat protection, while others make scientific and cultural contributions. Restorative development will keep spreading as the long-term economic value of intact ecosystems is demonstrated to local inhabitants, businesses, and governments.

WHAT YOU CAN DO

Individuals can personally contribute to biodiversity preservation in many ways. A simple plant, bird, fish, or animal naturalism book will help you become familiar with your local flora and fauna. Intimacy is the first step toward preservation. Planting a garden or even some window boxes with interesting and threatened plants is something many biodiversity gardeners are now doing. Heirloom species are gaining in popularity among home gardeners. Support your regional botanical gardens, and talk to the local "green thumbs" about what they know. Much preservation is carried out person to person. The Resource Section lists several great institutions actively involved in plant and animal preservation. It also lists other groups that are making a real difference worldwide. Above all, get involved, learn all you can, and experience biodiversity directly with a walk in nature.

In the end, only a profound philosophical shift in how we view our relationship to the natural world can assure that we halt our plunge into a biologically barren future. Ultimately we must ask the question: Does the entire natural world belong to our species? As we have the power to re-shape the biosphere and induce mass extinctions, what is our responsibility as the "dominant" species in this finite ecosystem? We hear a lot about the "right to life," but what about the "right of life" for all species? Spiritual and religious leaders around the world are increasingly raising these profound questions in their communities (see Chapter 8).

This chapter focuses on Vandana Shiva, a scientist and woman whose voice is lifted eloquently in the key of biodiversity. She goes beyond conventional scientific rationales into a heartfelt knowing of the genuine kinship of all life. Hearing her passion stirs our hearts to know that we are indeed all connected in the "Sahayak," or democracy of life, as people in India view it. As this bioneer so imaginatively illustrates, the gifts of biological diversity reveal profound spiritual and cultural dimensions that offer us an extraordinary opportunity to deepen our connection to the grandeur and spirit of nature, and to learn our own nature.

Vandana Shiva and the Vision of the Native Seed

Vandana Shiva has participated in a large social movement in India to preserve native agricultural seed stocks and traditional farming practices. Her work has led to the creation of community seed banks, empowering farmers to retain control over India's seed heritage and food supply. Her work opposing the Green Revolution and corporate control has revived ancient farming practices which are sustainable. She has challenged corporate plant patenting and established a spiritual basis for biodiversity conservation. The model is spreading throughout lesser developed countries and influencing the burgeoning seed movements in developed countries.

VANDANA SHIVA FELT LIKE A SLEUTH SOLVING A MYSTERY. SHE HAD BEEN traveling frequently to conduct forestry studies in the lush state of Karnatika in South India, where the famous Chipko movement had migrated from the Himalayas. *Chipko* literally means "tree huggers," describing the hundreds of local women who had gained international notoriety for bravely embracing the trees threatened by the jagged metal teeth of axes and saws. Large corporate interests were intent on uprooting diverse natural forests to replace them with monocultures, huge fast-growing plantations of single species for the commercial market. Now a drought had struck the country. Cattle were dying, and people didn't have enough food. But somehow the data didn't add up. Why, Shiva puzzled, was India spending the most drought relief money in Karnatika, the state with the highest rainfall?

Shiva discovered that the area had recently been converted to the production of hybrid sorghum. The new hybrid seeds, the core of the "Green Revolution" overtaking India, were characteristic of the new "miracle" plants. Designed to produce very high yields, the plants were bred to be short in order to take up as much chemical fertilizer as possible to promote fast growth and prevent them from falling over under the weight of the

48

oversized seed heads. The hybrid sorghum had already proved very vulnerable to pests, and now it balked at the drought, even though it was not severe in the region.

Shiva surmised that the problem was biomass scarcity. The hybrid plants left no organic matter to replenish the soil, nor did they provide essential fodder for the livestock integral to Indian agriculture. The basic biological cycle of decomposition and nutrient replenishment was failing.

"DAUGHTER, GET ME ONE SEED"

Despair and fear rippled through the ancient agricultural community. "A very old farmer said to me," Shiva recalls with shivers, "that if our old seeds were used, there would be no drought. He said, 'Daughter, get me one seed of our old sorghum, and I will drive the drought away from this district.' That's the day in my heart I made a commitment to seed saving."

Shiva was correct in identifying a biomass scarcity resulting from the newly engineered Green Revolution plants. They starved the soil and were vulnerable to disease and slight environmental variations. In the native seed, she saw the restorative cycle, the ability to adapt through the endless changes of nature. The incident planted in her the seed of a new path. With her field research team, she began to scour the countryside for old seeds. Several years later, they laid hands on the very sorghum seed the wizened farmer had pleaded for. Shiva, an eloquent, highly educated physicist, found herself in the tumultuous birth of one of the most important biological and social movements of the late twentieth century: the conservation of the world's biological diversity of agricultural seeds and the accompanying protection of her nation's food security.

But Shiva also felt herself deeply conflicted between her formal scientific training and her cultural heritage. She had often traveled as a child with her father, a senior forest conservator in the wooded foothills of the Himalayan Mountains. She dwelled in the forests and never saw the city until she was 15. "For us, entertainment meant finding a new form of life, seeing a new flower," she recalls, radiant in the embroidered folds of her red and gold sari. "The beauty of nature was the substitute for the disco."

Shiva's mother was a highly educated senior education officer who turned to farming after the partition of Pakistan. From scratch, she started a farm of her own, where young Vandana spent many holidays. Her mother was also a writer and often wrote about ecology. At a time when India was undergoing "modernization," building huge dams and calling them "modern temples," her mother was writing about the forest as the model of existence. While expecting her daughter, the pregnant farmer wrote extensively about forests and species. "In India," Shiva says brightly, "we have these lovely stories about how part of your education takes place when you are in the womb."

Shiva worked on the farm, which was adjacent to a prominent wildlife reserve. Knowing nature was her passion. It informed her growing interest in physics as the "most foundational study of nature." Because girls were discouraged in school from taking science and mathematics, the young woman took outside courses, and she eventually earned her Ph.D. in quantum theory in Canada. "My life split in two." She chuckles roundly. "There was the part of me that loved nature in its fullness and diversity. Then there was the intellectual approach to nature through the discipline of physics. It is a tragedy that we are literally a century behind some of the new sciences. We are still handling things as if the world is the projectiles and cannonballs of Newtonian physics. Quantum theory gives such a rich idea about how the world functions, and it is much more appropriate to living ecologically in nature."

TWO KINDS OF BIONEERS

Quantum theory deepened Shiva's vision of an age of biology, distinguishing a fundamental conflict of scientific models. "There are two kinds of bioneers," she says. "One will look very much like the pioneers who thought that every land they conquered was an empty land with no people and no prior inhabitants. There was no need to respect any rights. On the other hand, ecological bioneers recognize that every step we take is in a full world populated by a tremendous variety of species and many other people.

"The empty-land ethic leads to violence against species and to genocide. The notion of limitlessness that comes with the colonizing mind assumes that there are no limits of nature to be respected, no ecological or ethical limits, no limits to the level of greed or accumulation, to inequality or the violence unleashed on other species and people. Ecologically, we know that limits form the first law. There are limits to the nutrient cycle, and the water cycle, limits set by the basic rights of diverse species to exist, limits on our actions if you respect other beings. There are ethical limits if we are to be human beings. Sustainability is built on limits."

According to the European colonists of North America, the empire of "mankind" was to be established over all inferior creatures, Shiva reminds us. But for the people of India, to the contrary, "The Earth family has been not just all humans of diverse societies, but all beings. The mountains and rivers are beings too. In Hindi, the words *Vasudhaiva Kutumbam* means 'Earth Family,' the democracy of all life, all the little beings and the big ones with no hierarchy because you have no idea ecologically how things fit in the web of life."

While at the university, Shiva was chosen for a fellowship for a prestigious Exceptional Scholar Program, and each summer she spent half her time on special courses and the other half trekking where she had once traveled with her father. There she became closely connected with the Chipko movement in the very region of her childhood. "That was to me the healing, the recovery," Shiva remembers with a sigh. "Seeing the same places I had seen as a child with lively streams and full forests become deforested regions under development projects, with water disappearing from the streams, I started to spend every summer working with the Chipko movement. It was nothing more than trekking, except now it was not from forest to forest, but from village to village. The entry point was not nature, but people. I was not going to be able to relate to the beauty of nature without seeing the potential of nature to meet human needs, and the injustice involved in destroying nature because it also destroys people's survival base. Ecology to me became a social issue and a political issue."

After completing her Ph.D., Shiva was receiving job offers in North America, but Mother India beckoned. "I was very close to the forests of India, but I was very insulated from the society that is India. I'd always wanted to understand more deeply the battles of why a country like India with such an amazing past has to see so much poverty today. Why, in spite of having the third largest scientific community, does science make no difference to the lives of our people?" Shiva chose to return to her homeland to build a bridge between science and society.

THE VIOLENCE OF THE GREEN REVOLUTION

There Shiva painfully encountered the ground-level destruction caused by the Green Revolution, whose promise of abundant food and global peace was turning out otherwise. The Green Revolution had originated after the shock of the peasant-based Communist takeover in China in 1949, as peasant revolutionary movements spread like a prairie fire through angry, impoverished Third World nations. Fearing global upheaval, the developed nations initiated a deliberate strategy to supply cheap, abundant food to prevent political unrest. The Green Revolution seeds were, however, part of a larger package, conditioned to grow only within the narrow tolerances of costly petrochemical fertilizers and pesticides. The program also required expensive heavy equipment and massive high-tech irrigation. While initially the "miracle high-yielding" seeds did produce bigger crops, this gain proved to be at the expense of the environment and small farmers.

Shiva started to document the impact of Green Revolution technology on India's agriculture, people, and ecology. In a country where 70 percent of the population is involved in peasant agriculture, environmental problems hit farming villages first and hardest. Shiva watched vast monocultures replace the local biodiversity of seed stocks. She saw formerly self-sufficient communities retooled to produce crops for an external market economy.

"The monocultures were depleting the biodiversity that is necessary for survival in a subsistence community," Shiva recalls with dismay. "It does not grow food just for selling. It is first and foremost a self-provisioning community. And all agricultural systems have had a necessary balance among cereals such as the staple of millets. We had sorghum, pearl millet, finger millet, foxtail millet, barnyard millet—a tremendous diversity. We had tremendous diversity of pulses, what we call dal or lentils. I love to stop in the market in Delhi where they will have six kinds of onions, twelve kinds of beans, and eight or nine kinds of oil cake. Those were the kinds of diverse production systems that are also absolutely essential to having good cattle, which provide animal energy for the farm and dairy products. "Soon hybrid wheat and rice replaced traditional diverse diets. Although India had been the largest producer and exporter of oil seeds in the world, virtually disappearing were oil seeds such as sesame, and oil cakes, a byproduct used for feeds for the animals central to both farm production and food.

The impacts were environmental, too. "Diversity in the field is also balanced nutrition for the soil," Shiva notes. "You get the nitrogen-fixing from leguminous crops, and by using mixtures and crop rotations, you don't have soil depletion. The records of alluvial India tell us that the yields had been the same for centuries before the Green Revolution."

According to Shiva, crop husbandry must produce for three kinds of food: food for the soil; food for our partners, the livestock; and food for people, both for consumption and for market. This whole interdependent life cycle was suddenly ruptured. Farmers were compelled to produce mainly for market and found themselves with less and less to eat as input costs rose dramatically with petroleum prices. "These were really prescriptions for making farmers poorer. The soil gets poorer. Biodiversity is eroded. There's poverty at every level, except for the corporations which make money selling inputs and commodities."

The Green Revolution, funded extensively through the Ford and Rockefeller Foundations, also revealed itself to be an extension of a larger financial pattern. Researching old tax records from the period of British colonialism in India, Shiva found startling evidence that prior to British rule, for every 1,000 units that a farmer produced, only 300 units left the village, of which only 50 went to the central authority. Soon after British rule, the ratio changed to 700 units leaving the village, with 500 going to the central authority. Now with the Green Revolution, virtually everything was being exported. According to her calculations, for every 1,000 units produced, 2,000 were actually flowing out, including the fertility of the soil and local nutrition. No debt was being recorded for nature's depletion, just the accounting shell game of "pseudo-surpluses" of higher yields for sale in markets. It was colonialism in a "green" cloak.

The immediate result was a quarter of farmers driven off the land. Only large farmers could survive, but after 20 years, even they were buckling under debt pressure. The outraged farmers formed a volatile antidebt movement, refusing to repay debts. Shiva witnessed the movement unifying farmers against the Green Revolution and the wealthy transnational corporations. She engaged in what she calls "participatory action research," and began a controversial book called *The Violence of the Green Revolution.*

FORTY THOUSAND YEARS OF FARMING AT RISK

Shiva was witnessing the burgeoning farmers' movement turn violent. In Punjab, India's largest state, Indira Gandhi attacked the Golden Temple during an ostensible war against terrorists and armed religious movements, but, according to Shiva, the real agenda was to suppress the large movement of farmers protesting government agricultural policy by refusing to supply grain to Delhi. The farmers were outraged over the entry into India of multinational agribusiness corporations driving them to bankruptcy.

"I wrote the book because of censorship," Shiva says evenly. "I used to sit every morning and do my newspaper clippings, and suddenly there were no clippings on Punjab, except to say, 'A bus blown up, a train blown up.' It was really a silencing of the farmers' outrage against a nonsustainable agricultural policy. This was in the most privileged part of India, where the most money had been spent on agriculture. Yet even in this wealthy pocket there was absolute discontent. At least three or four times when I planned trips to Punjab, I would have to cancel because the train had gotten blown up. When I finished the book, publishers wouldn't pick it up. It was part of a censorship policy. So I published it myself. It sold out in no time."

In the 1991 book, she referred to the experience of Sir Albert Howard in the early part of the century when he was sent by England as consul to advise India on improving its agriculture. After years of study, Howard concluded that the British had little to teach India, but had a great deal to learn from Indian farmers who had successfully maintained the same rich farm land for 40,000 years. Shiva also corrected the misleading language applied to the "high-yielding" seed varieties, instead labeling them as "high-responsive" varieties since their yields depended on massive synthetic fertilizer use. She further documented certain traditional Indian nonhybrid seeds which gave yields equal to the hybrids without outside synthetic chemicals.

Shiva's analysis has since borne out. The Green Revolution led to "no change in the growth rate of Indian agriculture in the last twenty years while depleting aquifers and destroying soil and soil fertility," according to the report "The 'Second India' Revisited" by the World Resources Institute.[4] India, whose 900 million people constitute one of every six people on Earth, had a 37 percent rate of undernourished households by 1988, up from 12 percent in 1978.

The entry of multinational agribusiness corporations into India in 1988 opened up large-scale commerce in the pesticide and fertilizer sectors. But by law, Green Revolution seeds were still dispensed solely through government subsidies, and the seed supply remained in farmers' hands. The image of the native sorghum seed had insistently stuck in Shiva's mind, when an ominous political trend forced the seed issue to the fore.

While attending a meeting in Switzerland on biotechnology and patent laws in the proposed General Agreement on Tariffs and Trade (GATT) regulations, Shiva became distraught after listening to representatives of the large agriculture-chemical-biotechnology conglomerate Ciba-Geigy. The company was outlining the need for intensified vertical integration of corporate agribusiness with chemicals, pharmaceuticals, and biotechnology for the company to survive as one of perhaps a dozen such giant corporations in the twenty-first century. A reporter queried the shell-shocked Shiva on what could be done in the face of such concentrated power. While responding to the overwhelming question, she experienced an epiphany.

THE VISION OF THE NATIVE SEED

"If the patents debate had not come up, the farming community would not have stopped seriously to think of how quickly they were losing the resource of seeds they had nurtured. GATT created a shock because they could look further down the road and realize how terrible things could be. They had to look at the current erosion of seeds." Indeed, the supreme advantage of hybrid seeds to the transnational corporations lies buried in patent law. Claimed as "novel inventions" by corporations, hybrid seeds could be patented worldwide according to GATT. As vertically integrated companies, they were seeking to control agriculture starting with the first link in the food chain, seeds. No one would be able to reuse, trade, or re-sell the patented seeds without expensive licenses from the owners. At stake was control of the food supply.

"The seed became a symbol," Shiva remembers of her cathartic insight. "Surely our people must have been feeling this way when the British ruled us. It's all so powerful. They have the military, they are brutal, they have no conscience, they shoot, they kill, they imprison. And then Gandhi said, 'Let's pull out the spinning wheel, and put it in everyone's hands. The wheel by itself is nothing, but put in the hands of millions of people with a will to change things, it becomes a powerful instrument.'" In a country where a widespread cottage-industry textile business was displaced by cen-tralized machine production, the spinning wheel became the symbol of popular empowerment. It reinvigorated the popular economy and the spirit of the nation, and led to the seemingly impossible overthrow of British colonialism.

"The mechanization of textiles was the first industrial revolution, and now we're in the third industrial revolution, the engineering of life. If for the textile revolution Gandhi picked up the spinning wheel, then it makes sense in the period of the engineering of life to pick up the seed. That's the day I felt a political possibility that in the seed was nature's diversity. Each bit of seed tells the story of the community. In the seed is a political state-ment of what kind of life we want, what kind of agriculture we want, what relationship with the soil we want. The seed is a wonderful metaphor of that from which life arises. Something so small gives rise to something so big."

Shiva's study of quantum theory had shown her the fallacy of the cer-tainty of prediction. Everything has potential, and potential may lead in diverse routes but will never have just one outcome. The uncertainty of evolving potentials is what lets the seed evolve into a unique tree. Free-dom, she realized, is keeping as many potentials open as are in the system. To the contrary, imposing determinism in a very uncertain world led to re-ductionist ideas that everything must be produced through centralized food systems, or that the only way to impose seed characteristics is through ge-netic engineering.

PATENTLY ABSURD

Shiva's vision of the native seed was synchronous with the flood of events overtaking Indian agriculture. The GATT Trade Related Intellectual Property Rights (TRIPS) regulations mandated that plant-patenting laws must be applied in so-called lesser developed countries, including India. The prize is a genetic gold rush in the patenting of plants and life forms from the natural world. Large agribusiness corporations moved instantly to patent seeds in India.

Cargill, a $47 billion private company which is the largest grain trader in the world, filed for patents on traditional Indian seeds. The patents would prevent farmers from saving their own seeds for planting or trading. Farmers would be compelled to buy the seeds at a price estimated at 60 times the existing cost in a nation where 80 percent of people are peasants. Farmers were incredulous that the company would claim that altering perhaps one or two genes out of the 23,000 to 100,000 found in a plant justified the seeds as the company's "invention" and therefore private property.

Infuriated Indian farmers viewed the seeds as their cultural property, or as the collective heritage of humankind. Rebelling, a group of farmers in 1993 went to the second floor of Cargill's administrative headquarters in Delhi, removed all the company's files to the street, and set them on fire. Several months later, the farmers razed Cargill's newly built $2.5 million seed-processing facility. Shortly afterwards, over half a million farmers, wearing the green shawl that became the emblem of the movement, amassed in Bangalore for a rally to protest the GATT plant-patenting laws. They called the movement a "seed satyagraha," *satyagraha* meaning "fighting for truth," the same charged word used by Gandhi to describe nonviolent resistance against colonialism.

Cargill soon announced that it would no longer patent seeds in India. The company made the astonishing statement that patenting hybrid seeds was not really necessary because hybrids lasted for only three years anyway before they needed to be replaced by another new variety, and it wouldn't be worth the expensive wait for patents. Cargill's extraordinary admission undercuts the credibility of the entire hybrid system. These chemical-dependent, vulnerable hybrids, which are supplanting the sturdy nonhybrid seeds tested in the field over hundreds and thousands of years through all the vagaries of nature, were admitted to be little more than a flavor-of-the-month club.

"Every time I think of the term *intellectual property rights*, it makes me mad," says Shiva of the patented seeds. "Rights to products of the mind is bad enough, but rights to life forms as products of the mind is what it really means. Every time I do workshops and try to translate the legal concepts, women in the villages say that this is mad. I call them 'intellectual piracy rights,' and there are two levels of piracy involved. The first is piracy from nature because we have the arrogance to say that a seed or an animal is merely a product of the mind. Any patent on life forms must steal from nature because it must deny nature's creativity.

"In the case of biodiversity, it's also very often a theft from the Third World. In the last fifty years, industrial society has become a chemical society based on substituting for nature. Yet the people on the margins in the Third World have continued to live on the gifts of biodiversity. So when a company wants to do prospecting for medicine or collect seeds for improved varieties, what they do is walk into a community in the Third World, ask them how they use their plants for healing, or ask them how the crops do. All you come back and do is tinker. Things like Peruvian natural cottons have been patented without even tinkering!"

Shiva cites the example of the neem tree, a national treasure of India with a long folk history of multitudinous virtues, such as use as a natural pesticide and as a medicine against many conditions including scabies and malaria. W. R. Grace Corporation, the large chemical and pharmaceutical company, studied the neem and patented it in 1994 as a biopesticide, causing outrage in India. On the poignant tenth anniversary of the Bhopal disaster, where a Union Carbide pesticide factory exploded, killing many people and poisoning thousands of others, Shiva went hunting for a neem during the monsoon and planted it in her backyard as a symbol of India's traditional pest control.

Indeed, while seed patenting began in the 1950s, the field expanded to the patenting of other life forms and natural products in the 1980s following the U.S. Supreme Court's Chakrabarty ruling allowing a patent on an oil-eating bacterium. Ironically the inventor admitted that he had invented nothing, but simply moved some genes around. But the ruling opened the gate wide for biotechnology patents.

In 1994, W. R. Grace received an amazingly broad patent for all transgenic varieties of cotton, so outrageous that it was contested by competing companies and overturned. The company has, however, been granted patents for all soybean genetic modification, and recently gained a monopoly right on embryo transfer. "This means all animal reproduction," Shiva states flatly. "We as women know that whatever is done to cows is soon done to us. It would sooner or later become a monopoly on human reproduction." Shiva points out that a patent was recently granted for relaxin, a hormone naturally produced by women during childbirth. "Giving birth to children could become piracy from the corporation that has the patent on relaxin," she speculates with indignation.

Shiva criticizes the patent system for privatizing what has been and should remain a biological commons, imposing private monopolies on a shared knowledge domain. In the area of seed patents, Shiva identifies the issue as one of human rights. Without free access to seeds, people cannot conduct agriculture, and thus they forfeit their sovereign right to feed themselves. After all, food security is the foundation of national security. "Our freedom struggle now is freedom from patents on life, from monopoly control on what is essential to the very survival of people, the basis of their health care and the free basis of their food system. We want freedom from all aspects of life being controlled by a handful of corporations for which the patents might be essential for their survival, but not for the survival of the planet or the people.

"My argument against intellectual property rights goes back to a beautiful song we have in India about Lord Krishna. He says to his mother, 'I did eat the butter, mother. But don't say I stole it, because how can food be stolen? Food belongs to everyone, and what belongs to everyone can't be stolen.'"

THE THEFT OF THE ARK

Shiva traces the instrument of patents to a long-standing political pattern of domination by commercial interests. "We shouldn't get too shocked that the root of discovery is ownership," Shiva points out. "The little piece of paper that kings and queens gave Columbus on behalf of the pope was called the 'letter patent,' which said, 'Go out, and if you find any land not ruled by white Christians, then conquer it on our behalf and rule it for us.' It's when the Europeans arrive that the place gets 'discovered.' That's exactly what's happening now with genetic reductionism. It's when the DNA language is put as a flag on an organism that the organisms are getting 'discovered.' That's exactly the way life is being treated now under the new intellectual property rights debates. It's saying, 'Go out and find a new life form, and if it hasn't been tested by another corporate scientist, go ahead and own it on our behalf.'"

Indeed, Shiva sees the Green Revolution now transmuting into the Gene Revolution in agriculture, food, and medicine. "When life has become a mine for genes, nothing is safe. That's really the basis for our resistance in India to the slave trade of the twentieth century, the trade in genetic materials." In India, the first transgenic patent was ironically for the redesigning of amaranth, known in India as "God's grain." Shiva characterizes amaranth as a "brilliant" grain that survives low rainfall and protects other plants against pests. It is a vegetable as well as a grain whose nutritious leaves are edible. A fistful of amaranth placed in a pot of millets will help them keep without spoiling for three years. "We are trying to conserve amaranth in its diversity," Shiva says with conviction.

SEEDING DIVERSITY

Cultivating diversity is the antidote to narrow-focus genetic manipulation, according to Shiva. She cites an ancient Indian conservation system, the method of 12 grains cultivated together, a polycultural conservation system known as *navdanya*, which means "nine seeds" to signify diversity. It functions in every major ritual, such as the nine seeds used in ceremonies of birth and death, and nine seeds signifying the planets. Navdanya is believed to reflect the balance of life in the fields and also the cosmic balance. "The obligation to cultivate diversity is a cosmic obligation to keep a larger balance in place. That's the way in which we have been undertaking seed conservation in India." Forsaking the reductionist rhetoric of "conserving genetic resources," Shiva instead calls the seed-saving program "navdanya."

"I remember a farmer working with me through a mixed cultivated field, and he said, 'Exactly as those plants are partners of one another and help one another, they are one another's *sahayak*. We are procreators with nature, but we could not be creative if the creativity weren't there in nature.' You go to the farmers, and they instantly say that there is no way we can accept the violence of intellectual property rights to our culture, our ethics, our economic status."

Shiva adds that corporations want only the benefits, not the responsibility for the risks of genetic tinkering. Companies maintain that there is no danger or effect from the release into the environment of herbicide-tolerant plants, which permit increased use of their pesticides. Yet recent data shows otherwise, as such transgenic plants mix with 50 percent of wild species nearby in the first generation, spreading the trait into the interconnected web of nature, where toxins enter the genetic life cycle and weeds develop resistance to herbicides. Research on modified oil-eating bacteria has shown that when the bacteria go into the soil, the soil's microbial bacteria are reduced by about 50 percent, impeding the natural composting cycle. "There is more and more proof," Shiva concludes, "that the genome of genetically modified organisms is highly unstable."

The seed satyagraha has resulted in the creation of community seed banks across India, with which Shiva has worked closely. With advice from seed experts, she helped create the local infrastructures for seed saving and trading. Ten community seed banks have now formed, and farmers are learning to become their own processors, suppliers, and marketers. Through a nonprofit organization, Shiva is helping farmers set up cooperative seed companies for nonpatented varieties. She is especially proud of the beautiful seed packs, which show a picture of Gandhi in his classic slippers exchanging seeds with a farmer.

"BIODIVERSITY IS THE ULTIMATE COMMONS"

What the country needs, Shiva believes, are patent-free zones, much like nuclear-free zones. "Biodiversity is the ultimate commons," Shiva says firmly. "We will not allow it be enclosed. The United States resisted getting agriculture into GATT in its early days because food security was considered a sovereign right. But now the agribusiness corporations have become so big that they control the U.S. government. But Article Three in the Intellectual Property Rights treaty of GATT does say that if there are serious objections and threat to the public order—to the moral order of society—then countries may exempt certain areas from patency. This is our opening to block the patenting of life."

Shiva sees global cooperation including from the United States and developed nations as critical to having an impact against this theft of the ark. In fact, significant movements are springing up in the U.S. and Europe among farmers and gardeners to preserve traditional seed stocks. Nonhybrid seeds are increasingly popular, and mounting attention is being paid to patenting laws.

Shiva warns that the issue of food safety, a specter of alarming dimensions with genetically engineered foods, must be decided by local communities, by individual consumers, by regions, or even by countries. Yet at this time, such issues are furtively overseen by the Codex Alimentarius, an international body on which all the transnationals sit. She also warns against the trend that those who protest genetic foods are being criminalized, such as companies that were sued by huge Monsanto, producer of genetically modified rBGH milk, (altered with engineered Bovine Growth Hormone), when they labeled their own milk as free from the engineered hormone.

Security, Shiva reminds, invoking quantum theory, is leaving room for uncertainty to play out. Among the surprises that have emerged is that many leaders of the farmers' movement are women, often destitute women abandoned by their husbands. They have shown the greatest initiative to conserve native seed.

For Shiva, the involvement of women is essential for success. "I think women must perforce lead more multidimensional lives than their male scientific counterparts do. Because women are mothers and wives, we are very often more members of the community. When things go wrong in the community, the women are the ones who get the signals first. India is a highly divided society in terms of men and women. Yet when it comes to public activity, a woman is far more accepted in India than in Western society. Traditional patriarchal men from rural communities accept me totally as a daughter or sister. The old farmers with whom I work call me 'daughter' or 'mother.' It is kinship."

But there is an even greater issue for Shiva. "To me, the fundamental driving force is the reverence for life. Without it, I don't think I could have the driving force for feeling the pain of a farmer who is going to be dispensable ten years down the line. It's respect for the beauty of that peasant's life that creates in me the political engagement. The reverence for the life of other human beings, especially those treated unjustly, is deeper because they stay more human than those who have treated them inhumanely."

At a large rally in October 1994, Shiva was moved to tears when peasant farmers gave her three large urns filled with coins donated in 25 and 50 rupee pieces, tiny amounts of change scrimped together by the impoverished people. The urns were so heavy that she could not lift them. The farmers wanted the money to go for an international institute on sustainable agriculture.

"That old farmer had once said to me, 'Get me an old seed,' and that germ he planted in my mind saw itself unfold and is now reaching maturity. That urn that the farmers delivered to me is a new responsibility which I have to deliver on, setting up really serious sustainable agriculture options. This alternative has to be a viable one that makes a difference to where society is going as a whole."

Shiva's hope for the future inevitably returns to quantum theory, to the reality that living systems are complex and unpredictable. "It is because the piracy is so big and the powers are so concentrated that actually every step we take becomes very big too. It multiplies the power of impact. In this age of biology, for every step of colonization, you need only one domain that is free to block the colonization, because we are talking about life which reproduces. When life reproduces, it multiplies, and so from something small, big things can grow. In biology"—Shiva grins—"what happens in one part of the organism starts affecting other parts, and organisms are part of a larger field. Nonlocality is just the notion of solidarity. Far away across continents, we could be doing things that change the evolution of the potential for someone else. The world is not just physical and material."

CHAPTER THREE

The Riches of Human Expression

"THE DIFFERENCE BETWEEN A COLONIST AND AN INDIAN IS THAT THE colonist wants to leave money for his children and that the Indians want to leave forests for their children," said a Colombian anthropologist.[1]

The extinction of traditional indigenous cultures around the world is as drastic as the loss of biodiversity. Cultural diversity and biodiversity are intimately linked, because traditional cultures have long preserved their delicate interdependence with rich, diverse ecosystems. Their profound Earth-based knowledge carries crucial lessons on how human beings can live in an extended harmony with nature.

Nearly every fruit and nut we use was first grown by an indigenous culture. More than half of the foods we eat today came from North American Indian tradition. The Incas grew almost as many types of crops as all the farmers of Europe and Asia combined, often by methods that have yet to be equaled in their productivity and sustainability.

Ethnobotanist Mark Plotkin recounts how the seemingly chaotic, overgrown gardens of Amazonian Indians actually reveal intricate marvels of biological variety, companion planting, and pest control carefully designed to provide maximum nutrition with minimal work and no damage to the forest.[2] By comparison, the attempts of desperately poor non-Indians to farm the rain forest with inappropriate slash-and-burn methods leads to failed crops, degraded land, and human misery.

Apart from losing the profound Earth wisdom of indigenous peoples, their extermination and the systematic erosion of their ancient cultures also represent a grievous violation of human rights. Protecting these human rights is essential to restoring the Earth.

Virtually all traditional cultures are embedded in a deep spiritual connection with the natural world. At the core is a respect for all life as sacred. This Earth-honoring relationship is a model that is serving many people around the world today in their own quest for spiritual and cultural renewal.

THE PROBLEM

The last few hundred years have been characterized by the invasion and plunder of indigenous people's resources by the "dominant" cultures. World trade has caused the systematic displacement and destruction of native peoples to satisfy a seemingly insatiable demand for goods such as spices, gold, fur, sugar, coffee, fruit, rubber, oil, cattle, and uranium. When native lands are found "useless" for all else, they often become toxic storage sites.

Most people today are aware of the history of genocide conducted against indigenous peoples. Within about one hundred years after the European encounter with the Americas, the Aztec population fell from 25 million to 1 million. Brazil alone lost half its tribes from 1900 to 1950, leaving about 200,000 Indians from an estimated 2.5 million in 1500. The North American native population in the current area of the United States fell from 1 million to virtual extinction in 400 years.[3]

This horrific loss, repeated around the world, has increased in the twentieth century with an even more effective systematic cultural annihilation. More than half the world's 15,000 indigenous languages have disappeared, and only 5 to 10 percent of the remainder are likely to survive another 50 years.[4] It is not necessary to slaughter people to destroy their culture. Severing their ties to their land, undermining their spiritual values, seducing their youth with consumer goods, and prohibiting their languages in schools have proven equally devastating.

While indigenous peoples occupied most of the planet only a few centuries ago, today their ancestral lands have been reduced to less than one quarter of the Earth's area. Many of these lands are the most rugged and remote, but also the most richly biodiverse regions on Earth. Technically, nation states offer them very few "legal" guarantees to the vast majority of it. Their few land rights are almost always conditional, excluding mineral, water, or other fundamental material rights. The African Pygmies, who have occupied their rain forest home for 40,000 years, have no legal rights to their ancestral land. Treaties that do exist are routinely flouted, undermined, or not enforced.

Indigenous people everywhere are usually at the bottom of the socioeconomic pyramid. Unemployment among many U.S. and Canadian tribes exceeds 50 percent. Native people are also frequently relegated to the riskiest occupations, such as working in uranium mines.

Observers have compared the loss of knowledge from the death of a single indigenous shaman to the burning of the ancient library at Alexandria. As economic pressures and cultural erosion intensify, young people are losing their link to the land and their ancient cultures. This vital wisdom is vanishing forever. Preserving the Earth wisdom and human rights of indigenous peoples is among the most urgent and compassionate struggles today on behalf of the Earth.

SOLUTIONS

People around the world are gaining a greater respect for the wisdom and Earth-based traditions of indigenous cultures. To create a future that works, the rich legacy of tribal peoples must play an essential part.

Indigenous peoples generally have an exemplary record of land stewardship. Most of the last surviving healthy ecosystems on Earth are cared for by tribal peoples, who have learned to integrate human needs with the diverse life of the land.

The Kpelle of Liberia plant over 100 varieties of rice, carefully matching crop strains to sunlight, moisture, slope and soil conditions in fields that are "jigsaw puzzles of genetic diversity."[5] The Shuar of the Ecuadorian Amazon use over 800 species of plants for medicine, food, fuel, shelter construction, and hunting and fishing tools. Some traditional Southeast Asian native herbalists use up to 6,500 medicinal plants in the same sophisticated manner as the highly developed classical Chinese and Ayurvedic medical systems. The grassland Sukuma of Africa rotate their grazing animals on cycles of 30 to 50 years. The Zaghawa of Niger drive their camels and sheep north in parallel paths and leave ungrazed strips for the return home.

Naturally, there are exceptions, as environments have also been degraded by tribal peoples. But most indigenous cultures also venerate the land and the cycles of the Earth, constantly giving thanks and acknowledging a spiritual power or Creator. Their Earth-based cosmologies are teaching many nonindigenous cultures around the world to reexamine the connection between human spirituality and our planet.

Many groups today are effectively working for the survival of indigenous cultures. International networks have formed that include indigenous peoples and human rights and environmental groups. The critical first step is guaranteeing indigenous land rights, and there has been some progress. Several nations now recognize substantial native land claims, although they rarely include full resource rights.

Another hopeful trend is the increasing number of "joint management" experiments all over the Third World. Government agencies, realizing they have no hope of monitoring immense resource-rich areas with very limited personnel and equipment, have begun working with local communities, often indigenous ones, to encourage them to act as stewards and protectors of their regions. There are many success stories of sustainable economic projects that earn income while preserving the integrity of indigenous land and culture.

Indigenous peoples all over the globe are resisting and organizing. In extreme cases, this struggle takes the form of armed resistance. More common are nonviolent protests, such as the Penan logging-road blockades in Malaysia and the Chipko forest movement in India. Indigenous peoples and their allies are effectively using national and international legal appeals and suits.

Cooperation among peoples is growing. Pan-tribal and international organizations are helping one another, printing newspapers, starting radio stations, and even, as the Kayapo in Brazil have done, videotaping politicians' promises. Their efforts are aided by a wide variety of sympathetic local, national, and international groups. These groups play a very valuable role, whether pressuring governments and lending institutions, or spotlighting reprehensible corporate behavior, or educating the public, helping with legal appeals, raising funds, and advising on sustainable development projects.

Any strategy to ensure the survival of indigenous peoples and the lands they inhabit must include the development of viable sustainable economic enterprises to provide income and raise community living standards. They also allow these communities the choice of modernizing and adapting to the global economy on their own terms, at their own pace, without jeopardizing their cultural identity or their environment.

One key for this process to be successful is that much indigenous knowledge be accorded the status of "intellectual property rights." While controversial, this arrangement allows communities to share legitimately in the profits of products originating from their lands or developed from plants used in their medical pharmacopeia. Alternatively, the struggle to abolish certain patents altogether is also raging, as Vandana Shiva's words in the last chapter attest.

WHAT YOU CAN DO

As one person, you can get involved in preserving cultural diversity by tracing your own roots, much as Francisco X. Alarcón describes in the following section. Traced back far enough, all world cultures were once indigenous, with most rooted in a reverence for land and life. Remembering and finding inspiration in that tradition is part of any solution to restoring the Earth. Cultures are never static, and we are in the dynamic throes of reinventing ourselves by making choices about how our species fits into the natural world. Remembering and renewing our personal and collective history is a profound first step.

In addition, read and learn and talk to people. Make friends with people from other cultures. There are many worthy groups working in the area of human rights and cultural diversity (see Resources Section) which you can join or even travel with. Your purchasing power can support companies and groups engaging in "fair trade" with traditional and indigenous peoples, since economic survival is a precondition of cultural survival. There are also many important efforts by groups to preserve indigenous lands, human rights, traditional cultures, languages, and shamanic practices that you can support and get involved with. But you can also tap into your own personal experience, create your own ceremonies, and feel the power of ancient wisdom, which is available to all who seek it. You may want to revise your own calendar to include Earth-honoring rituals such as summer and winter solstice or days that were important in the Earth culture of your own ancestors.

Such a quest can take great vision and courage, as the following profile describes. As a mestizo, a blend of European and Indian extraction, Francisco X. Alarcón has probed deeply into the painful contradictions inherent in his mestizo heritage and the violent history of Mexico and the Americas. He has returned with sublime poetry invoking this legacy, facing the future with new hope. As Alarcón portrays, many of us are of mixed heritage, and we are all capable of the process of "reindigenizing" ourselves toward a world culture for the next century. After all, culture is a human creation, and we are all recreating ourselves all the time.

Francisco X. Alarcón
Rediscovers the Americas

Francisco X. Alarcón is a mestizo poet who has used language as a magical door to understand the multicultural heritage of humanity. As a person of mixed Indian and Hispanic lineage, he believes that his personal experience represents a mirror of the global cultural flux that will become a bridge to respecting the important legacies carried by the many cultures of the world, including indigenous ones. All cultures, he believes, carry a strong Earth-honoring tradition, and we can all reconnect with our roots to create a future world culture based on respect for the land and one another.

ACCORDING TO THEIR CUSTOM, FELLOW CHICANO POETS LUCHA CORPI, JUAN Pablo Gutiérrez, and Francisco X. Alarcón convened on New Year's Day in 1989 at Juan Pablo's apartment in San Francisco's Mission District. "This day I have a special gift for you, Francisco," Lucha said with a mysterious smile. "What kind of gift?" asked Francisco at the pleasant surprise. "Lie down on the floor, and I'm going to give you something," she replied. "No, no, no," he said warily. "I'm not into this." But seeing the anguish in her face, he understood that it was something important to her, though he couldn't imagine what. "Relax," said his friend, "and let your mind go."

Lying on the floor with his dark black ponytail splayed across the bright Mexican rug, Francisco felt Lucha's hands moving above his body. Suddenly, he began to experience a vision swimming before his lidded eyes. "She became my grandmother. The vision was like a dream, but I could tell the dream to my friends at the same time." Juan Pablo wrote it all down.

A VISION OF ANCESTORS

"I was dealing with a manuscript," Alarcón recalls with passionate intensity of the life-changing experience, "that was written in Nahuatl, the ancient

71

language of the Aztecs. In the vision was this tropical place in another time. Lucha asked me, 'What is your name?' I said, 'My name is Yo,' but that was not my name. The experience lasted about fifteen minutes, and it was very magical. I asked her, 'What is the meaning of this?' She told me, 'It's your problem now!' She was relieved. She is a visionary and has certain abilities. She was an angel to me, a gift of the gods. For weeks I didn't know the meaning of this."

Alarcón began to read avidly about vision quests. That summer, guided by intuition, he felt called to travel to Mexico, where he had spent part of his childhood. He went to Mexico City's famous El Museo Nacional de Antropología e Historía (National Museum of Anthropology and History). "It was as if I knew just what to do. I had written a letter that I was a researcher. The staff there didn't want to help me because they saw that this is a Chicano who doesn't know Nahuatl. I insisted for a week, and they finally came to the conclusion that I was not going to leave until they gave me access to the manuscripts in Nahuatl."

In the Mesoamerican collection Alarcón found what he didn't know he was seeking. It was a rare manuscript written by one Hernando Ruiz de Alarcón, a Catholic priest born in Mexico who had been commissioned by the Spanish Inquisition to record the Nahuatl magic spells and religion a hundred years after the conquest of the Americas. "The authorities were concerned," Alarcón says with irony, "that the Indians were still worshipping the trees, and still thinking the lakes and the streams were sacred. This had to be stamped out."

Alarcón's distant relative from long ago had interviewed about 50 native informants on their culture, religion and belief system, one of the first ethnographies of Mesoamerica. The bilingual author spent ten years writing the book, completed in 1629 but never published and kept hidden for 350 years. Only in the nineteenth century did someone find and publish it, although only an obscure 200 copies exclusively seen by scholars. "I was totally taken with this collection," says Alarcón. "I had to smell the manuscript, touch the ink, feel the spirit of the informants. It really gathers the belief system of the Nahuatl people in this part of central Mexico and the southwestern United States too.

"Hernando Ruiz de Alarcón was a very knowledgeable person, but he was doing this book in order to destroy this belief system because it was somehow against the Catholic faith. But, by recording the Nahuatl spells in the original language and translating them into Spanish, what he actually did was to maintain the tradition. I believe now that maybe he himself became involved in the tradition, because he spent too much time explaining in detail how to say the spells, how to divine with corn, for example. The spells are incredible, and they are not from the temple tradition. Part of an oral tradition of the common people dating back 3,000 years to the Olmecs, they are about how to live every day, about hunting, about farming. How to greet the sun in the morning. How to talk to the trees so they can give wood. How to get rid of unruly ants. How to find affection, and how to heal with herbs. These are still our concerns."

At the time, there was tremendous turmoil within the Aztec empire when the sedentary peoples of the south were conquered by the nomadic, warlike people of the north. The culture confronted an ethical dilemma about human sacrifices, and many people were reverting to the older tradition of Quetzalcoatl, the mythic Feathered Serpent from whom life had originated and who presided over the cosmic balance. According to legend, Quetzalcoatl left the Toltec city of Tollan in 911 C.E. because he opposed human sacrifice. The people were thereafter divided in profound soul-searching over the loss of their god's favor. Their prophesy bespoke the eventual return of Quetzalcoatl, which they awaited expectantly.

THE MAGICAL DOOR OF LANGUAGE

As a poet, Alarcón became entranced by the Nahuatl language, which 3 million people in Mexico still speak today. He was especially taken with a phrase that appears in all the spells: *nomacta nehuatl.* Although it is so different conceptually from English or Spanish that it can be translated only loosely, it means "I myself." It establishes the position of the speaker. Alarcón characterizes it as a "shamanistic incantation, like 'abrakadabra.' It means that the subject and the universe are one," he continues in the whisper of a storyteller at a campfire under a full moon. "It's a magical space where the person says, 'I am the spoon and the coffee, and the table and the sofa, and the roof and the building.'"

Similar phrases convey the overarching concept of unity where subject and object become one. *NOxomoco, niCipactonal, niQuetzalcoatl, niNahualteuctli* means "I myself, I the first man, I the first woman, I the feathered serpent, I the enchanter." "It's not a metaphor," Alarcón says with elation. "It's not a poetic image. It is that you *are* the thing. The Nahuatl language reflects another way of relating to the universe."

For Alarcón, his 1989 encounter with the lost manuscript came some 500 years after the beginning of the European conquest of the Americas, sending tingles through his cellular memory. As a contemporary Chicano writing poetry, he did not want to undertake a conventional anthropological study. Instead he chose to engage in a poetic encounter. "I wanted this book written in my own voice, the voice of my grandmother, the voice of my ancestors. I was thirty-eight years old, and it had taken me this long even to see myself in the mirror. It's almost a miracle that now I can see the high cheek bones and Olmec mouth, the dark skin of my hands. My experience is that of millions of people in the Americas, Africa, and Asia. We are the outcome of the expansion of the West, and of the tremendous sacrifice that the native peoples of those continents have had to pay for that expansion. 1992 was an important date for us to reclaim our identity. By reclaiming our past, we reclaim our future."

In the book that grew from his searching, *Snake Poems: An Aztec Invocation*,[6] he wrote:

*Mestizo**

my name	behind
is not	my Roman
Francisco	nose
there is	there is
an Arab	a Phoenician
within me	smiling
who prays	my eyes
three times	still see
each day	Sevilla

*All poetry in this chapter is from *Snake Poems: An Aztec Invocation*, by Francisco X. Alarcón, published by Chronicle Books, San Francisco, 1992.

but	my feet
my mouth	recognize
is Olmec	no border
my dark	no rule
hands are	no code
Toltec	no lord
my cheekbones	for this
fierce	wanderer's
Chichimec	heart

THE MATRIARCH CALLS
THE SPIRITS IN NAHUATL

For Alarcón, the personal journey of discovering his identity became a mirror for understanding society's encounters of cultures and races. Born in Los Angeles in 1954, Alarcón knew that generations of his family had washed back and forth across the Mexican border with the regularity of tides, but tides turned by human hands. His relatives had first come to the United States in 1911 after the Mexican Revolution, when 10 percent of Mexicans moved north to escape the violent upheaval in their home-land. His great-grandmother and grandmother lived in Texas for eight years before coming to fabled California in 1919. His great-grandmother never returned to Mexico or learned English, but remained otherwise very traditional.

Alarcón's grandparents were married and had five children when the Great Depression struck in 1929. "To be Mexican in Los Angeles at this time was almost a crime," he says painfully. "So in 1931 my grandparents were forced to go to Mexico because my grandfather had lost his job as a mechanic because he was Mexican. This was the end of his American Dream. He never came back to the States and was very resentful because he had believed in the system. It was a tremendous crisis for my family. My uncles were already in high school here. They were born and raised here in the States, and suddenly they were in Mexico. This is my own conflict, belonging to two cultures. The border is right here inside me."

Later, Alarcón's mother returned to Los Angeles, where Francisco X. Alarcón was born. Then the family migrated once again to Guadalajara, where Francisco grew up. It was there that he came to know his father's mother, a very traditional Mexican and speaker of Nahuatl who largely raised the boy. Her influence was profound.

"My grandmother had a tremendous ability for communication and was always telling stories," he remembers with reverent affection. "Everything had meaning for her. I would go to her house, and it was always as if you were going to have a lesson that day. In December, she would offer *tunas*, the prickly pear pads of the cactus, to the Virgin of Guadalupe. I love *tunas*, and I always wanted to eat them, but I never dared do that because they were part of an altar. I have learned that the word *tuna* comes from the word *tonal*, which comes from the word *tonalli*, which means 'spirits.' It's also the name of the sun, *tonatiuh*. So my grandmother was calling the spirits by using the Nahuatl term. Her universe was very organized. To me that is wisdom.

"My family were peasants in Mexico," he recalls, "and I got to know a lifestyle that was almost biblical. There were no appliances, no electricity, no roads. People lived in the mountains and would tell stories at the end of the work day. A lot of people in Mexico today cannot survive that way because they're bombarded with radio and TV. How can *mestizos* survive in that way? How can I as a writer, a person who now lives in the States, provide for this culture to continue? My writing actually started when I was thirteen by trying to retrieve the memory of my grandmother and her stories and poems. She died when I was eighteen."

Matriarch

My dark
grandmother

would brush
her long hair

even ferns
would bow

to her splendor
and her power

WE ARE ALL CONNECTED

After attending an elite Mexican Jesuit high school as a scholarship student, Alarcón began college at California State University in Long Beach while his family was still in Mexico. Working as a dishwasher to get by, he gained a degree in history, ironically in classical studies of Greece and Rome. But he found his cultural identity dogging him on both sides of the border. "It didn't make any difference if I was in Guadalajara studying with upper class Mexicans or at Stanford. They all had a disdain for mestizos and dark-skinned people. I have very dark skin and am the most Indian-like of my family." As a child, he had been tagged as "El Indio," a pejorative mockery of his native bloodlines.

Alarcón moved to the Mission District, a largely Chicano neighborhood in San Francisco, where he founded a magazine, *El Tecolote Literario* (The Literary Owl). "We wanted to challenge the Anglo-American establishment. The idea was to promote diversity." He helped establish the Mission Culture Center, one of the largest Latino centers in the country. He completed his Ph.D. in Latin American literature at Stanford, crafting his writing, which had become very important to him.

But the Mission District proved to be his "true university and the setting for a search for roots. Here I am, a person who comes from a peasant community to this fool's modern postindustrial society. Somehow in my own lifetime I moved from working in the fields in a close relationship with nature to an industrialized modern society to an elitist U.S. university and living in the Mission. I decided that what I wanted to do in my life is to be *one*. My grandmother provided me with a role model for doing so."

Delving into Ruiz de Alarcón's obscure tract provided the raft he sought to return to the shores of his lost identity. Studying his ancestor's story in the Nahuatl tongue began to transform his anger at the horror of the deliberate extermination of a culture. He adopted language as a means of reconciling the internal split that was tearing him in half.

Hernando Ruiz de Alarcón
(1587–1646)

It was you
you were looking for
Hernando

searching
every house
every corner

for some
dusty seeds
of *ololiuhqui*

it was you
who you tricked
and apprehended

it was you
who both questioned
and responded

everywhere
you saw Moors
with long knives

and in front of
so much sorrow
so much death

you became
a conquered
conqueror

priest
dreamer
speaking cross

condemning
you saved yourself
by transcribing

maybe
without knowing
the heavens

I am
from your tree
from your dream

this *cenzotle* bird
in the wilderness
your tomorrow

For Alarcón, unraveling the *mestizo* experience became a path for seeking the integration of inner conflicts. "I'm a *mestizo* because I'm in turmoil. Somebody called my poems haikus, but I said they're like tattoos because they puncture your flesh. We're not in peace. We see ourselves in this tremendous wound of history. Every time I look at myself, I see rape and the violation of our ancestors. I teach Spanish at the university, but Spanish is not my real language. I have to search for that language, and for the last six years I've been conjugating Nahuatl in my mind. Language is a door to talk to my ancestors."

As Alarcón probed his poetic encounter with Nahuatl and his literary ancestor, he began to see ancient Mesoamerican patterns emerge from beneath the layers of conquest, like paint peeling off an icon to reveal an earlier hand at work. "Chicanos call their friends *cuate*, a slang word which comes from *coatl*. It actually means 'snake,' or 'my double.' In the Mesoamerican concept, time is a spiral represented by two snakes. Each year a snake sheds its skin, and a rattlesnake grows a new ring, marking time. Time is a spiral, and the future is going back to the past. The snake is a very sacred symbol in Mesoamerica, such as the feathered serpent Quetzalcoatl representing circular time and fertility. So *cuate* means that time goes in a circle, and we are all connected."

The idea of duality is also deeply rooted in Mesoamerican thought, Alarcón found. Genesis in the Mesoamerican tradition is held in the union of two spirits, *Ome*, which means 'double,' and *Teotl*, which means 'divine force.' When Ometecutli and Omecihuatl came together, they made love and begat the universe. "To me, it's like the Big Bang theory," says the poet. "It is a balance of male, active energy and female, receptive energy. Making love is divine, and sex is nothing to be ashamed of."

CORN IS THE SUSTENANCE
OF THE UNIVERSE

The mythology of corn is also intertwined with the Mesoamerican story of genesis, which tells of several appearances and disappearances of human beings. Quetzalcoatl as a god of culture was very lonely since an earlier world had been destroyed because the people had lost their way. He wanted to invent humankind again, and he did it through corn. But there were certain gods who did not want humanity back again, so Quetzalcoatl turned himself into a little ant, went underground, and tricked the Corn Monkey, who was guarding seeds of corn. He stole some corn seeds and two bones and brought them back to the surface. He pricked his penis and let the blood run over the corn and bones. From this ritual, humanity emerged again to live on the Earth.

"Corn is the sustenance of the universe," Alarcón observes with pleasure. "Corn was known as *Tonacacihuatl*, 'Our Lady of the Flesh.' In the 1960s some anthropologists in Veracruz, Mexico, found a local community that was using the old incantations, but making them to the Virgin of Guadalupe. For me the Virgin of Guadalupe is the corn goddess. She's the Mother of Christ, but she's also 'Our Lady of the Flesh,' which is corn."

For Storing Corn
Ruiz de Alarcón (III.5)

I myself
Spirit in Flesh

come forth
elder sister
Lady of Our Flesh

soon I shall place you
inside my jade jar

hold up the four directions
don't shame yourself

you shall be my breath
you shall be my cure

for me, Poor Orphan
for me, *Centeotl*

you, my elder sister
you, *Tonacacihuatl*

Just as corn is sacred, food is a sacrament in Mesoamerican culture, Alarcón has come to understand. "Traditional Mexican food has a mythology. I eat tortillas every day of my life. I came across Nahuatl writings about the tortillas that were called *tlaxcalli*. *Tlaxcallis* actually represent the sun. So when you're eating the tortilla, you're paying homage to the sun. What a beautiful way of praying! Every time I roll a tortilla, I will be rolling Father Sun. And I will eat in communion with the sun."

A POETIC ENCOUNTER WITH THE AMERICAS

While immersing himself in the poetry of Nahuatl in the summer of 1989, Alarcón again decided to follow an inexplicable intuition. He felt called to visit the easternmost point of Mexico because he sensed that a hidden cycle was somehow linked to this historic place where Conquistador Francisco Hernández de Cordoba had first arrived in 1517 to begin the colonization of Mesoamerica. The Spanish chronicler of the conquest, Bernal Díaz de Castillo, wrote, "We sailed from Havana. Twenty-one days after we had left port, we sighted land, which made us happy and for which we gave thanks to God. This land had never been discovered nor was there any knowledge of it till then." The expedition came in search of slaves and gold.

Díaz was describing the tiny island of Isla Mujeres, the Island of Women, in the turquoise Caribbean off the coast of the shimmering tropical Yucatán peninsula in eastern Mexico. Today the island is a favorite tourist resort in an uneasy peace with the old mariner community.

Alarcón walked to the eastern shore of Isla Mujeres despite admonitions from locals that there was nothing to see there. Beneath a lighthouse near a cliff dropping off into the sea, he found the unmarked ruins of a temple. It was the temple of Ixchel, the mother goddess of healing in the Mayan Indian tradition. A sacred site considered by the Mayans to be an oracle, Ixchel fell quickly beneath the Spanish sword, as the invaders destroyed the temple, took women as slaves, and ransacked the temple for its decorative gold.

"This was the beginning of the nightmare," Alarcón says, his voice choking with emotion. "I asked myself why the Mexicans and mestizos have not recognized this place as part of our history. I was there for almost a whole day crying. It was a catharsis, because I was Francisco coming on the boat and I was a native looking at Francisco. I had landed and I raped and took off, but I was one of the women too. I was both invader and victim. But I was also the wave, the stone, the star in the distance."

An anguished Alarcón walked back to his hotel in the descending darkness. He wept for the slaughter of Mesoamericans whose numbers plummeted from 25 million to 1 million in the first 100 years of the conquest. "We have to reclaim our suppressed tongues and spirits," he wrote, "our burned homes and fields, our slaughtered mothers and fathers, our enslaved sisters and brothers. By reclaiming ourselves, we will reclaim America." For Alarcón, America is not the narrow strip of North America that is the United States. It is a land without borders that stretches from Alaska to Patagonia, and Americans are all the peoples who have lived in the hemisphere.

"Somehow we have this separation in the West. We have body and soul, self and nature, and a mythology based on an ego that is individual, that is not connected with anything else. Thus we can abuse and destroy nature. The Mesoamerican way is different in that we are all connected. You see the connection between animals, plants, hills, stones. We are all part of one, and birds speak and have feelings, and stones have spirits also. That's not the same in the Judeo-Christian Bible, where animals have no soul. I think that explains why we have the nightmare we have today. Somehow we have to wake up from this nightmare, and maybe the mestizo people will have to wake up and see themselves in the mirror.

"Earth-based religions and belief systems are still here. In the West they have been repressed, and thousands of people have died because they believed that the Earth is sacred. That was a tremendous crime. The point now is for us to reemerge with nature-based religions. The native traditions of Mesoamerica and North America are so important today precisely because they are able to maintain this relationship with nature.

"I feel very optimistic for the future," he continues softly. "I see an America coming that will be a more humane America, a reconciliation. Paradise is here, now, today. That's why the Nahuatl tradition is so appealing to me. The physical is the spiritual. The political is the spiritual. There's no separation. When you call the spirits for divining, you say 'Nomatca nehuatl, NOxomoco, niCipactonal': 'I myself, the first man; I myself, the first woman.' We are the first and the last. I recognize myself as a complete being who is connected with the wholeness."

But for Alarcón, the transformative power of the Nahuatl language holds the key to restoring the ancient culture. "Language is a key of culture. A culture is memories preserved of a relationship with the universe. A language establishes that culture. It's very intimate, but it's very collective. It's crucial for native languages to be maintained."

Having made the commitment to a poetic encounter, Alarcón was astonished to find that other scholars did not see the Nahuatl incantations preserved by Ruiz de Alarcón as poetry. "The spells recorded by Hernando Ruiz de Alarcón were poems," he says incredulously, rolling his eyes heavenward. "In Nahuatl, *xochitl* and *cuicatl* are two words which mean 'flower' and 'song.' Together they are the word for 'poetry.'"

In Xochitl In Cuicatl

every tree
a brother
every hill
a pyramid
a holy spot

every body
a seashore
a memory
at once lost
and found

every valley
a poem
in *xochitl*
in *cuicatl*
flower and song

we all together:
fireflies
in the night
dreaming up
the cosmos

every cloud
a prayer
every rain
drop
a miracle

"People go to Mexico"—Alarcón chuckles good-naturedly—"and they ask, 'Where are the Indians?' I say in return, 'Who are the servants? Who are the people doing the dishes? What happened to the people? They're here!'

"Demographically, America is becoming more like me. You'll see a lot of cheekbones like mine here in America. We are here. We've been here for thousands of years. The future is going back to the past. I say the future of America will be mestizo and that all the cultures will come together. When we are able to come to an understanding that we are all in the circle, life will be fine."

But for Alarcón, poetry is the heart of the mestizo rebirth. "As a poet, I'm not the author of the collection. The author is the whole collective voice of the people that has survived for thousands of years. We are all mestizos."

And so Francisco X. Alarcón continues to invoke the Nahuatl tongue in a new ecopoetics to restore the sacred. It is a Nahuatl belief that sacred corn is the happy grain, and he hopes that the presence of corn in much of our foods (it is present as an ingredient in 70 percent of foods) will weave the desired spell and heal our separation from the Earth and from one another.

Potent Seeds

few corn
kernels
enough

to turn
anger
around

CHAPTER FOUR

Greening Medicine

THE CEDAR FORESTS NEAR THE BASE OF MOUNT FUJI, A GLOBALLY RECOGNIZABLE symbol of Japan, are strikingly beautiful in the green cloak they cast over the gently curving landscape. But by springtime, heavy with amber buds, the trees exude a fog of pollen which afflicts one in ten Japanese. Tokyo displays two giant electronic billboards blinking with the latest pollen counts, but the pollinosis score doesn't alleviate the miserable sneezing and mass allergies created by this most unnatural disaster.

The mountains formerly hosted a diversity of native trees until after World War II, when Japanese government foresters replaced them with vast monocultures of cedars. The fast-growing cedars seemed a smart industrial choice, a semi-hard wood with predictable uniformity suited to large-scale lumber production. About 40 percent of Japan's forests were replanted with commercial trees, almost half of these being cedar.

Now the people sniffle unhappily at the unprecedented concentration of trees whose pollen clouds are so thick that they mimic forest fires. The designer forest has also threatened wildlife, precipitated serious soil erosion, lowered the water table, and enhanced the slippery prospect of mudslides. It has also failed economically, incurring $30 billion of debt through subsidies. Even the government forest engineers admit that disrupting the natural balance of the diversity of trees was not such a clever idea after all, for the forests or the people.[1]

The state of the environment is arguably the single most important influence on our personal health. As we have degraded the environment, we are experiencing the dramatic increase of negative health impacts. And it is through personal health issues that growing numbers of people are awakening to the need for environmental protection. Very large numbers of

85

people are responding by embracing natural medicine and a healthy diet, including organic foods and botanical remedies. The health benefits gained from these natural products are highlighting the fact that the Earth is often our primary source of medicines and healing.

THE PROBLEM

The increasing toxicity of our environment appears to be a growing factor in causing disease and illness. Rises in neurotoxicity, cancer, asthma, reproductive disorders, and the "new" categories of environmental illness and "sick building syndrome" are increasingly attributable to the 70,000 industrial and agricultural chemicals percolating through our collective environmental body.

Meanwhile, a host of devastating new diseases are appearing—AIDS, chronic fatigue syndrome, and Lyme disease—while supposedly eradicated scourges such as tuberculosis, cholera, and pertussis are returning. The destruction of tropical forests and previously undisturbed habitats is releasing new lethal viruses such as Ebola and others for which we have no name. The copious use of antibiotics is breeding resistance to these drugs as the pathogens adapt quickly. Genetic manipulation of the food supply also poses great potential public health hazards which are poorly understood, yet the government is readily permitting the development and sale of such foods.

Conventional medicine excels at the management of emergency medicine, trauma, and certain bacterial infections, but it has lagged badly in disease prevention. It has performed poorly at treating the vast majority of chronic illnesses, such as cancer and heart disease, and until very recently had almost totally neglected the link between chronic illness and environmental factors and diet.

Conventional medicine, formally known as allopathic ("against disease") medicine, was founded on the historical philosophy of "heroic medicine." This approach stipulates that the physician must intervene aggressively to "cure" the malady, generally using poisons and often dangerous, invasive technologies. Until the twentieth century, allopathic medicine was but one approach among many accepted medical traditions.

Medicine was diverse and pluralistic; it encompassed many natural approaches with reputable lineages, such as herbalism and homeopathy. Medicine had not previously been a profession to achieve wealth, and most doctors were poor or of very modest means.

The twentieth-century industrialization of medicine forever changed the economics and politics of health care, sweeping aside natural medicine and radically driving up costs. Today the U.S. health-care system gobbles up 14 percent of the Gross National Product (GNP), the highest in the developed world. The pharmaceutical industry boasts the highest return on investment of any industry at over 20 percent a year. Ironically, many of these pharmaceutical companies are the same conglomerates that produce the toxic fertilizers, pesticides, and industrial chemicals that cause the very health problems that industrial medicine then purports to treat.

Treatment of chronic disease currently accounts for 85 percent of the national health-care bill. While virtually nothing is spent on prevention, over 60 percent of all costs are spent on 10 percent of patients in the last three months of life — vast sums on heroic interventions for doomed patients.[2] Although the surgeon general's office has attributed 60 percent of disease to diet-related causes, official nutritional guidelines are either very unsophisticated or manipulated by outright vested interests of the food industry. Doctors receive virtually no training in nutrition in medical schools.

The bottom-line costs of treating illness are already mandating that we change our ways. The next century seems destined to place primary emphasis on preventing illness and engendering wellness. This trend inevitably will lead both to preventing pollution and to a much greater reliance on natural medicines that come directly from the Earth. In any case, green medicine is a clear example where large numbers of people have created a new pathway that is already changing the face of medicine and our relationship to the natural world.

SOLUTIONS

In the face of mounting medical costs and widespread dissatisfaction with results, a quiet consumer revolution has been taking place, an enormous grassroots defection from industrial medicine. An astonishing study,

"Unconventional Medicine in the United States" published in the *New England Journal of Medicine* in 1993, came to the conclusion that more Americans paid visits to providers of "unconventional therapy" than to all U.S. primary health-care physicians. To do so, people were willing to spend $13.7 billion annually on alternative therapies, three-quarters of which was paid out of pocket! As of 1990, one in three people in the U.S. was using an alternative therapy, a figure that may be even higher today.

In direct contrast to allopathic views of healing, natural medicine posits the existence of the *Vix Medicatrix Naturae*, a healing force of nature. From this perspective, the natural healer seeks to stimulate the body's own healing powers. Allopathic medicine today, though it still rejects this basic idea of a vital life force, has acknowledged the existence of this condition as the "immune system."

This popular natural medicine movement is now starting to make inroads in medical schools, clinics, pharmacies, and other mainstream channels, including some elective classes and programs at Columbia, Harvard, and Yale. At its core is a pharmacopeia of natural products, especially plants, herbs, and foods. In fact, about 80 percent of the world's population still relies on traditional herbal medicine for primary health care. This trend is now expanding dramatically in the United States and Europe.

A few large insurance companies such as the Oxford Health Plans are now adopting alternative therapy networks in a bottom-line acknowledgment that these therapies are generally safe, cost-effective, and popular. Oxford estimated that about one-third of its clients had used an alternative therapy in the past two years.

Natural products still comprise the foundation of modern pharmacy. As Harvard biologist Edward O. Wilson points out, "Few are aware of how much we already depend on wild organisms for medicine. In the United States a quarter of all prescriptions dispensed by pharmacies are substances extracted from plants. Another 13 percent come from microorganisms and three percent more from animals, for a total of over 40 percent that are organism-derived."[3]

As Wilson further notes, organisms are far more sophisticated than the world's chemists at synthesizing organic molecules of practical value. The saliva of leeches contains an anticoagulant that dissolves the blood clots that jeopardize skin transplants. The saliva of a Central and South American vampire bat prevents heart attacks. In 1990 Merck sold $375 million worth of anticholesterol Mevacor, derived from a fungus.[4] The return to natural products is being embraced by huge commercial entities. Pfizer, another giant firm, collaborated on a $2 million a year, three-year effort with the New York Botanical Garden to study plants in the U.S. with potential medical uses.

The rediscovery of "folkloric" medicine is gaining momentum and respect, especially as it relates to diet. It is now well documented that humble garlic, known as "Russian penicillin," offers, among several functions, a valuable medicine for lowering cholesterol associated with heart disease. Lactobacillus acidophilus, a bacterium found in yogurt, is another folk remedy now documented as a viable medical treatment for yeast infections and deadly infant diarrhea. Cruciferous vegetables such as broccoli and cauliflower are now known to contain strong anticancer nutrients, as do many common fruits and vegetables, which can be used preventively in our diet.

Natural medicine has been extensively codified by major civilizations such as the Chinese, Tibetan, and Indian Ayurvedic. In the West, advanced systems of herbalism, homeopathy, and naturopathy have survived to regain widespread public favor. Much of this knowledge originated in the vast reservoir of empirical experience of indigenous peoples. The random screening of plant extracts for anticancer activity gives 10 percent positive results, whereas when screened by ethnobotanical use, from 20 to 50 percent show positive activity.[5]

WHAT YOU CAN DO

As individuals, millions of people in the United States and around the world are taking personal responsibility for their own health and well-being by educating themselves and changing their lifestyles and medical choices. Diet is widely believed to be a primary determinant of our health, and revising your diet to include fresh, organic, and locally grown foods is well within most people's reach. Exercise, stress reduction, and the use of herbal and dietary supplements can also improve your health and prevent illness. If your doctor is not familiar with the kinds of natural medicine you want to institute in your regimen, then find one who is. There are now insurance plans that will cover natural medicine, or you could consider a high-deductible policy for emergencies and use the money you save on premiums for alternative care. There are many good books and alternative medicine groups that can help you find your way through the field of green medicine. Don't neglect to learn about the place you live, and understand how it is affecting your health. How can you help make it a healthier place? Or should you consider moving if it is a deleterious zone?

Many more people and companies today are recognizing the close link between personal and environmental health. Adding to that motivation to restore the Earth is the fact that nature's pharmacopeia still provides the most sophisticated array of health products for human well-being with far greater safety than synthetic drugs. Clearly the future holds a collaborative model that integrates conventional and complementary health care.

As plant explorers Kat Harrison and Steven King demonstrate in this chapter, the healing power of nature is a very good motivation to protect the environment. Harrison reveals a realm of *medicina* that broadens our understanding of healing into a universe of spirit and wholeness. Meanwhile, King's work with Shaman Pharmaceuticals is helping redirect the business of pharmaceutical health care back to the tropical forests, which provide some of the richest sources of healing plants. Both these bioneers inevitably point us back to the Earth.

Kat Harrison and the Plant Spirit

Kat Harrison is a botanist and artist devoted to exploring the spiritual nature of medicinal plants. She has conducted plant collection and biodiversity preservation efforts especially directed at tropical species with medicinal and shamanic traditions. She has also gathered traditional folklore associated with plants and the place of human beings in the web of life. She extends the idea of medicina to include spirituality and the healing nature of nature.

THE COMPUTER HUMS QUIETLY, FORSAKEN IN THE SMALL TREE-HOUSE-LIKE office of Botanical Dimensions. Its screen displays an article on a Peruvian botanical garden for the next *PlantWise* newsletter. Nearby are stacks of fundraising letters and volumes of botanical books. Colorful original paintings of the Amazonian jungle shimmer as though alive on the clean white walls.

Proprietor Kat Harrison has wandered next door to the house, ascending the steep overgrown pathway through old northern California redwoods. Big-leafed semitropical potted plants crowd the steps up to the house where she has perched for hours in the moist sunroom. The delicate flower of her peyote cactus is experiencing its periodic bloom, and Harrison has decided to witness the small biological miracle by sketching the fine-leafed pink petals and sunny yellow core. Peyote, the sacred visionary cactus of many North American Indians, who use it for spiritual guidance and religious ceremony, is one of her favorite shamanic plants.

"I made myself stop working in the office," Harrison says softly, her long sylvan brown hair catching the sun's gold. "It was blossoming for eight hours, and I sat down to draw it as realistically as I possibly could. I saw a whole new level of detail, even though I've drawn it before. It's just perception. Respect is the basis for biodiversity, and perception is the basis for having that respect.

"We have to open our eyes and really pay attention," says the smiling botanist, her liquid brown eyes scanning the cactus for details she may have missed. "I think that much of the root of humanity's long attempt to conquer nature is that once you stop perceiving it, you start fearing it. That's why you should draw. You've got to be able to see it to draw it."

For Harrison, being in nature is second nature. The slender, gentle artist grew up on Santa Catalina Island off the coast of California, without a car and with little contact with the mainland. Her naturalist father took the family to Mexico when she was six, and she recalls deep memories of immersion in nature, riding horses in the jungle, playing with iguanas, and walking naked in the fecund wilderness. She reveled in her mother's large gardens and began painting nature.

"I knew at that time," Harrison remembers, "that nature was the most important thing that had happened to me so far. If you're in a redwood forest, sitting straight, meditating with your spine against the trunk, you tune in and feel the strength of what the redwood is drawing from the Earth. You feel the force of that pump of light that's pushing up through that huge, ancient tree, and just become absorbed in it. It ceases to be an idea. It ceases to be something you value just because you love nature. You're really in it and feeling it."

After leaving high school and the island, Harrison started college and then took up traveling, spending the better part of the next ten years roaming the Middle East, North Africa, Europe, and then Mexico and Latin America, which drew her back repeatedly. "I would ask people about local methods of healing and about their dreams. I would ask old people for stories. I loved communicating with people who came from a simpler background than I did, for somehow we had a similar heart. I loved waking up in the morning and not knowing what was going to happen."

IN SEARCH OF THE PLANT SPIRIT

Meanwhile, Harrison studied botanical and biological illustration in school. "I kept training my visual perception to see if I could see what this essence was." It was also her introduction to seeds and scientific nomenclature, differentiating the formalized classification of the diversity and variety of plants and life forms she was learning. "I really got into plants through the art of seeing," she reflects. "But I found another level ultimately. You really begin to see the spirit of the plant."

In 1976 Harrison spent three months in the Peruvian Amazon with partner and then husband Terence McKenna, an ethnobotanical explorer. The mission was to find Ayahuasca, a visionary vine sacred to native peoples of the region. Working with a local *curandero*, or medicine person, Harrison became immersed in their way of seeing, which she characterizes as a "hypersensitivity to the interrelationship between people and plants. It was the cellular-level perception that everything is vibrantly alive in the same way that we feel we are. Every organism has its unique qualities. As a species, it has a vast set of relationships to all the others in its world and to its ecosystem.

"I remember someone saying to me that if you harm this plant, all the other plants like that—even the ones a thousand miles away—won't like you anymore. Each species has a history of relationships to other species. It has allies. It has enemies. It has jealousies and dreams. It is a whole being. You're really having a relationship among species, and need to perceive them with that kind of respect."

Harrison found herself becoming a "plant person," a term she coined. "It's like a wedding of people and plants. Once you become a plant person and you've really gotten how great plants are, you can't stop. Hawaii is populated by plant people, like the ancient little Japanese ladies who always have their clippers in their purse. You go everywhere with your plastic bags, labels, and clippers for the rest of your life. You're always trading. It really is like the birds carrying the seeds caught on their feathers. We're doing that part of evolution, which is moving the species around and having them meet one another."

During her foray into the Peruvian rain forest, Harrison met Manuel Cordova-Rios, the "Wizard of the Upper Amazon," a supreme plant person, *curandero*, and shaman. He expressed to her his concern that the plant knowledge held by indigenous people was rapidly being lost. He advised her that the real treasure in the herbs in the marketplace lay in the mind of the person selling them. His warning led Harrison and McKenna to found Botanical Dimensions in 1985. She later gained an M.A. degree from Sonoma State University in ethnobotany through interdisciplinary studies in biology and anthropology.

The work of Botanical Dimensions (BD) has focused on collecting not only living plants and seeds but also the plant lore from cultures practicing folk medicine in the tropics. The nonprofit enterprise places special emphasis on medicinal and shamanic plants, and supports collection efforts in local areas. BD maintains a singular sanctuary in a private botanical garden in Hawaii where many rare plants are transferred and cultivated. The Hawaii effort is a "gene garden" and a "living library" whose forest mountain paths are planted as closely as possible to their original Amazonian conditions.

"YOU DON'T COME TO THE JUNGLE UNLESS THE PLANTS CALL YOU"

Along with the plants, Harrison has assiduously collected and assembled a computer database of the folklore from the plants' caretakers, usually native peoples. Her work more recently, however, has involved supporting collection efforts near Iquitos, Peru, by mestizo people of Indian and Spanish lineage. "I must have appeared an odd benefactor because I could never afford to get there till last year, although I had been sending them money for years. I could see in them flickers of these different races hybridized together into the people who were remembering their Earth wisdom and plant wisdom from a variety of traditions. They have a feeling of our all being brothers and sisters, and there's a real openness with the information because clearly we all need to know it now."

The garden is located in the upper Amazon at the edge of the Andes in a secondary growth forest. The lush setting extends into the heart of the jungle, populated by many indigenous tribes. The area contains one of the richest troves of biodiversity anywhere in the world. The caretakers have begun intensive transplantings of many rare and important plants to the garden.

"We spent a week just moving through every day as in a dream," Harrison recalls with a flush. "Something is ripe—let's go harvest it so we can eat it. Something's flowering—let's go see it. What do you want to know about? Let's go way out there and show you the thing you never even conceived of. Just all paying attention to nature together, a group of people out in the forest. It gives people a foundation to respect their own traditions."

The garden is called El Jardín Botanico Sachamama Etnobotánico, named after the forest mother vine, *sachamama*. The garden guardians go off on long trips to *los tribus*, the tribes, where they collect and bring back the plants and the folklore. She found a very different concept of medicine. "I began to see the whole arc of their kinds of medicine, or *medicina* as they call it. *Medicina* is actually any plant that brings us benefit. Rather than the Western model of the magic-bullet medicine for the identified ailment, in this other worldview, medicine really begins with food. The arc then goes on to tonics – plants which provide preventative effects— and then on to medicines of the kind that we're more familiar with medically: herbs to tone the liver or strengthen the heart. Then it goes into toxins, because plant poisons are very important and can be medicine to attack a problem or cleanse the body. Or they may be used as a teacher in shamanic apprenticeships, including the more clearly visionary plants such as Ayahuasca."

Along with garden founder Francisco Montes and three other men who were budding plant people, Harrison met La Abuelita, a grandmother who was the mother of one of the men, and who took care of a 19-year-old niece and a young adopted orphan girl. Their lifestyle was decidedly different. "I had been there a couple of days when I observed how they pay attention, how someone would notice a particularly beautiful leaf and everyone would stop to look at it. Or a frog would croak and the conversation would stop as though asking who had spoken. Or a bird would call and they would laugh in delight. I was getting this feeling of the depth to which they are embedded in a worldview of nature in which we are just one small part. In that moment I caught a gleam in La Abuelita's eyes, and it hit me so strongly that they really live in a world that is magic and everybody knows it. In our better moments, maybe we all know that the world is magic, but knowing and acting that way in one's daily life is really the incredible thing. I felt as if they were children in a garden."

Harrison was equally intriguing to the local people. La Abuelita observed that you don't come to the jungle alone, especially as a woman, unless the plants have called you. "She asked me what I wanted to know, why I had come. Spontaneously I said I had come really to learn about the plant spirits. For the first time I realized that this was the most important thing to me."

Harrison's hosts quickly set about helping her fulfill her mission. The women suggested a ceremony, leading her to a beautiful stream. La Abuelita picked all the aromatic plants she could find and made a gigantic pot of tea in a pink plastic washtub, which she steeped for the whole day in the sun streaming into the forest. The intoxicating scent wafted continuously up the hill to the settlement. At dusk they went back down to the river where La Abuelita performed a ritual for her honored guest. "I kneeled naked above the stream. I was thinking to myself, 'High mosquitoes, high malaria!' But trust, trust, trust! La Abuelita, who is at once very elfin and very ancient, poured this thick, thick tea over me until I was coated in a paste of plant material and just smelled so amazing. Mosquitoes didn't bother me—I must have smelled too good!

"It was intoxicating and, of course, tobacco is always involved in sacred ceremonies, so she was blowing tobacco everywhere and singing all these *icaros*, these magical songs. She's old, and she knows many of them. The songs felt charged with an aura of protection."

"EVERY SPECIES HAS A SONG"

"Every species has a song. If you are granted the song in a vision state, or by just submitting yourself to the presence of a plant and opening up, then it's a real gift. You are to remember that song forever and share it when it seems appropriate. That song has healing power, and there are some which are handed down from one *curandero* or *curandera* to the next, and there are others which come to us as individuals. But they are part of an encyclopedia on the sonic level of the same thing that the seeds represent on another level.

"I was coated with this mush, and I just lightly scraped some of it off, leaving a lot of stuff in my hair and on my skin. For a couple of days people came up to me just to smell me several times a day. What she was giving me was blessings."

ALL YOU HAVE TO DO IS DREAM

When she first arrived, Harrison was quickly attracted by a plant called *labios de la sirena*, the lips of the siren. When she touched it, her hosts considered it a sign that it was possibly her ally. "I have always had a thing about sirens," Harrison revealed to the attentive group. "I grew up on an island, and my father was a mariner. I knew the story of Odysseus and the sirens."

They proposed several experiments. "'If you take it with a piece of the *sachamama* vine, it allows you to hear the siren of the waters singing sad songs,' they said. It was suggested to me that I do their way of instant bio-assay of a plant (to ascertain its properties). You would take a few leaves and crumble them and rub them on your face and your forehead. Then lie down in the hammock or sit in the forest with your eyes closed and become very receptive, suspend judgment, be clear, and find whatever occurs. It is the unbidden information that's actually valued the most. That which we don't lead, that which we don't direct."

They experimented on plants using Harrison as their medium. Seeing her affinity for *labios de la serena*, they suggested that she put it under her pillow at night. When she awoke in the morning, six people were waiting to hear her dreams.

"I had amazing dreams, really the most beautiful, consistently positive, long, and totally memorable dreams I've ever had in my life, and I'm a good dreamer. Two nights in a row I had stunningly positive, glittering jeweled dreams with beautiful women and children and vast feasts. There was a beautiful party in a huge white house on top of a hill with sun and rainbows and snow women in beautiful twirling dresses. In the dream somebody delivered a letter to me with calligraphy of my name on the outside. I opened the letter, and a song came out, floating into the air. I would like to dream like that every night."

One day at the garden her hosts told Harrison about leaves that glowed on the floor of the forest. No one could identify them. They could be seen only in the dark, and even the men were wary of venturing into the jungle at night. But they wanted very much to know what the glowing was. What spirit was this glowing?

With only her tiny flashlight, Harrison embarked with the three men and the young woman into the moonless nocturnal forest. "Eventually we saw these glows in the distance and we found a few and then we found more. They were big leaves, two different kinds, a standard leaf and one spearlike leaf. They were bright and damp on the forest floor. Because they were damp and big, we could put them on our clothes, so we put them all over us while we were moving in the night."

The luminescent forms glided through the jungle blackness back to the sanctuary of the garden, where the hosts asked Harrison to put the leaves under her pillow and dream. "By now I'm getting orders every day. Put this one under your pillow and go to sleep. Hurry up! What are you staying up for? So I went back and put it under my pillow that night. I still had the *labios de la sirena* under my pillow. I had a terrible nightmare—really, really grotesque, one of the worst nightmares I've ever had. It was a far future world, all ice, spikes of ice like metal spikes, and there was ice coming off everything. All the children were missing, my children, everybody else's children—it was that level of panic. You couldn't move because of the spikes on everything. I woke up from it and I thought, 'This is so bad,' but I had promised them that I would do this experiment, and I didn't feel that I could take the leaves out from under my pillow."

Going back to sleep, Harrison awoke with another chilling nightmare. Removing the leaves, she saw that their luminescence was fading, and thought perhaps the plant was dying quickly. She placed the leaves by her bedside and fell back to sleep to have other exquisitely positive dreams.

In the morning she recounted the dread of her nightmares. Her hosts listened intently and noted that there were two species of leaves that had the luminescence. They conjectured that the spiney species of leaf was jealous. It was a very territorial plant spirit, they conjectured. Because it was so very territorial, perhaps it had seen that Harrison had such a strong ally in the *labios de la sirena* and became hostile.

"BEADS ON A NET"

"These plants are allies to each other, and that's part of the magic of this world. They're also *not* all allies. They can be enemies to one another. It's a network of activity and population. The point is to see how they are relating to one another. So when you get six plant people living in a couple of thatched huts in the middle of the jungle, you are down in the ocean of all of these forces. It's a net of species, and we're each just a bead on the net. Human beings are one bead in this net of the millions of species that are communicating. Take away your identifying mind and just see the spirit in the plants.

"Every single species, however humble it is, has eyes with which it sees the world, and it sees as legitimate and true a version of the world as what we see. That really puts a different light on our holding the control here. We seem to have the power to hold control, but there is another level of communication that these people know. There's a way of being open to the communication among all species. In these traditions, the spirit resides in the species, and every species, wherever you meet its member, will speak to you in the same way and from the same source."

For these mestizo guardians of the garden, reestablishing the link to nature is compelling. When the botanists go visit *los tribus*, they come back with rich stories, and often spend dinner telling them to one another to preserve the memories. Sometimes in the middle of a story, Harrison found, they will burst into a song, because that is the way to tell that part of the story. Together they construct herbarium specimens consisting of pressed plant parts, seeds, sometimes even the pollinating insects. They make herbal tinctures, medicinal resins, and incense. They talk constantly about plants.

Many interesting plants are coming to light in their investigations. *Una de gato* is a potent herb found very effective in a recent cholera epidemic. It is also widely used as an immune stimulator and antitumor agent. *Lengua pincha* is one that people take when you have to see an attorney. It can be used in any situation involving having judgment passed on you.

Their view of *medicina* extends beyond the question of what something is used for. "To appreciate their view," Harrison says, "to be cured by the plant, you have to know the whole family structure of the healing plant. There are certain songs that have to be sung when you take the plants. Maybe there is something behaviorally in your life that is exacerbating the physical problem you're having. That's where the *curandero* comes in, who might get a bad feeling coming from a relationship in your life. He may advise you to take the herbs when the moon changes, and sing a song into a little bottle of flower water, which you dab on your body at certain times for the next month. All these different things work together for healing. You have to be in balance and harmony in the world. You have to pay attention. You have to know the names of things. You have to know the stories. You have to honor the people who work with these energies."

Harrison believes that the mestizo people such as those at Sachamama represent an important model for the twenty-first century. "Because they are not directly from an indigenous tradition, although they have mothers and grandfathers who were native to the area, they have a lot of the tradition, but with many others mixed in. They have values that embrace and preserve what's around them, yet they ride downtown on a moped if they need to. They have one foot in the so-called real world, and one in the spirit world. We can all understand the kind of transformation this represents with the sad loss of traditional cultures, but with some kind of rebirth into a new way of recognizing one another: the real global society."

For Harrison, the solutions lie in creating garden preserves such as Sachamama for both the plants and the lore. "Because of the fragility of the whole picture, I just wish that there were a thousand awe-inspired botanists fanning out over the whole world to every fragile culture and ecosystem, writing everything down. Botanical Dimensions is about education. Our Hawaii garden is a respectable example of what can be done by individuals on a small scale. I've heard of a half-dozen places in Latin America made up of individuals collecting. Often they'll find one family of plants they're particularly fond of. We need more regional gardens based on folk wisdom."

THE VOICES OF THE ANCESTORS
SPEAK IN THE SEEDS

The new garden guardians believe that the plants may not endure without the guidance of the ancestors. For indigenous peoples, the plant spirits are often considered a pathway to the ancestors, who are held as close in death as in life. "There are plant-based religions throughout the world," Harrison observes, "and there are ritual traditions that go with these. The shaman's job is described as recognizing the gateway to death, being able to traverse that passage and return and bring back what is learned. These plant-based religions use this bridge for the voices of the ancestors to come through.

100

"There are many traditions in which they say the ancestors have a design already. If you listen to them, you can address your questions to them, and the design will be made visible. Maria Sabina, a well-known Mazatec shaman from Mexico, often referred in her chants to 'the path of the tracks of the palms of your hands.' In that tradition, sickness comes from your spirit leaving your body, when spirit has somehow gotten bumped sideways. So in order to backtrack, the healer has to go back to the moment when the spirit left the body, which they do by saying, 'We follow the path of the tracks of the palms of your hands.' It's your story.

"The seeds themselves—whatever species, wild or cultivated—are encoded with the DNA story, and each seed is a long, winding, subtle story. The seeds have crossed human hands and been cultivated by us, and selected by us and bred out and traded across continents and oceans, and saved from extinction at the last minute, or lost. These are really the voices of the ancestors speaking in each of those seeds. We don't have the seeds and species without the intervention of all the people that came before for thousands of years.

"We are the ancestors of the next stage. When we talk about the next five hundred years hopefully being better on this continent, what we're setting up is our role as ancestors for our children, those children five hundred years down the line for whom we will be those distant little voices in the seeds. 'The paths of the tracks of the palms of your hands.'"

Curious about the linkage between seed diversity and the legacy of ancestors, Harrison referred to a dictionary to find the classical meaning of *heirloom*, a word applied to old seeds handed down through human generations. *Heir* indicates to inherit something passed down from ancestors. In old English, *loom* meant implement, or something useful, and by the time of Middle English it meant the loom that you wove on. "So here you have something that we've inherited to weave with. With heirloom seeds, perhaps we are taking them and weaving the future, certainly the future of botany, but perhaps also the whole ecological awareness being born in the keyhole of history right now."

Harrison believes the planetary mission now is to go back on the species level to where humanity's relationship with nature became separated from spirit. "It's too late to be indigenous again, yet we need to settle into a deeper awareness of nature and place. Working with biodiversity is one way to do that.

101

"I had a perception once a few years ago that just above our heads is a river of magic. It's not a one- or two-dimensional river—it's everywhere. Maybe a so-called army of us could go out and be collecting information and trading seeds and propagating and fostering them. That's what connects me to La Abuelita, this tiny little woman in the Amazon. She and I both know about the same place. We both know that if we reached above our heads, there's a place of energy which is universal where you can shift things around. We need to make a spell that works on us most of all, the spell of changing our consciousness and changing our deep bad habits. That level of magic, which is accessible through the plant teachers, is where that kind of thing can happen."

The peyote flower has closed with the dusk, sealing its cycle for another season. The newsletter and funding proposals are stirring on the desk in the soft evening breeze, summoning. Kat Harrison glides down the stairs under the giant redwood to get back to work, guided by the plant spirit.

Steven King and
the Medicine of the Forest

*Dr. Steven King is helping transform medicine by identifying and
developing pharmaceutical drugs from higher plants based on indigenous
knowledge. With Shaman Pharmaceuticals, he is creating models of
sustainable plant harvesting and cultivation, while working to support
native people to preserve the lands and Earth wisdom of their cultures.
By adding a commercial value to biodiversity through innovative medical
development, King believes the rain forests and their peoples stand a better
chance of survival, while the world will gain invaluable medicines.*

WHEN THE 13-YEAR-OLD BOY CURLED UP IN THE CHAIR TO READ *KEEP THE
River on Your Right* by Tobias Schneebaum, the real-life adventure story
transported him down the great Amazon River with an explorer who got
lost in the jungle and then was found by the native people of the forest. It
was well after dark when young Steven King turned the last spellbinding
page, disoriented to find himself not in the teeming rain forests of Brazil,
but in concrete-and-steel Chicago. He resolved to alter his geography as
soon as he could. Sometimes geography is destiny.

A BUDDING BOTANIST FOLLOWS HIS PASSION

King soon happened into a synchronous series of opportunities to live out
his dream of traveling to the rain forest. Attending the College of the Atlantic in 1974, he landed an internship through a Peruvian anthropologist
from the Center for Amazonian Anthropology, who took him to study
with the Secoya Indians on Colombia's Santa Maria River. Returning be-
latedly to college from the thrilling expedition, King quickly located another anthropologist working in Peru, and wrangled his way to live with a
tribe and work on a study of the local plants, diet, and health. He collected

plant specimens and chronicled the native people's diet. "I was an entertaining nuisance," King recalls, stretching his long frame against the confining office chair at Shaman Pharmaceuticals' California headquarters, "and therefore I was tolerated and embraced. That experience changed my life. I was overwhelmed with the intimacy of their knowledge of their environment and plants. I was taken by the integration of their culture with the natural environment, and how well they managed to thrive in this wild jungle."

Looking for an expert on Peruvian plants, King met Timothy Plowman, a plant person who became a lifelong mentor. "A lot of people talk about this stuff," the skeptical botanist told him, "but few do it." He advised King to focus on plants, both medicinal and food, and King heeded the advice. Completing college, he linked with Calvin Sperling, a key player at the U.S. Department of Agriculture in plant and seed conservation, for a trip to the highlands of Peru and then through Bolivia, where he immersed himself in the food and medicinal plants of the Quechua civilization, one of the most advanced plant societies in history.

King later became a primary collaborator with the Smithsonian Institution's national exhibit called Seeds of Change, showing the profound contributions of the plants of the Americas. Italian food as we know it, King says with pleasure, did not exist before such native foods as tomatoes, potatoes, and peppers traveled from the Americas to Europe. The real wealth of the Americas was not silver or gold, but corn, chile, and blueberries, not to mention chocolate, tobacco, and sugar.

With characteristic serendipity, King also encountered Michael Balick, head of the New York Botanical Garden, just when the respected plant expert was founding the Institute of Economic Botany, which was designed to document the economic value of plants, strengthen plant collecting, and support native cultures. Balick offered the young man one of the institute's first full fellowships, At the time, botany was regarded as an archaic profession, a dried flower pressed between the musty pages of nineteenth-century science. For pioneers like Balick and King, this green world held an endless fascination for the mystery of the human-plant connection. It would also prove to be the leading edge in a global rediscovery of plant medicine increasingly led by economic incentives.

Suddenly the budding botanist was being paid to follow his heart. "I had no idea that any of the things I love to do would pay the rent. I was certainly told by many friends along the way who are lawyers or professionals that this was all entertaining and neat, but someday I'd have to grow up and get a life, a job, a career. My advice to people with a passion is to follow your passion, because you never know what's going to come down the pike."

HIGH-TECH SHAMAN

But Dr. Steven King would hardly have speculated that his formerly unglamorous passion would lead him to work with one of the more exciting companies in the world today, doing what he loves best at a time when plant medicine is experiencing a heroic renaissance. Striding briskly on long legs through the innovative company's high-tech laboratories in South San Francisco, King speaks in bursts of enthusiasm and staccato bytes about the science of plant medicine.

The lean, hawk-eyed botanist explains that the state-of-the-art laboratories are deconstructing massive amounts of chemical and biological data to create pharmaceutical drugs. White-coated technicians pore over complex machines, which drip a dark liquid brew to convert it into analytical chemistry readouts. Behind the glass-walled clean rooms, piles of dried plants are bunched in preparation for extraction procedures, a vivid contrast between the funk of nature and the dust-free high technology of contemporary biology and natural product chemistry. The plants are the difference here, King explains, because Shaman Pharmaceuticals is using higher plants as the sole source for its medicines. The company, of which King is today Vice President of Ethnobotany and Conservation, is also working closely with the native peoples who know these plants so intimately, whose knowledge is the cultural compass for directing the complex and otherwise blind drug discovery process.

Ordinarily drug companies use a process of random collection of plants and natural products to find new medicines. This hit-or-miss process is one of the reasons that the costs of finding a new drug have soared to around $325 million, a sum so gargantuan that only huge companies can participate in this most lucrative business in the world. Of the 250,000 higher plants known in the world, a mere 10 percent have been examined, and only 1 percent have been tested for medical activity. About half to two-thirds of those plants reside in the rain forests, which is one reason that the destruction of tropical rain forests is considered such a disaster.

Plant medicine, King points out, is the cornerstone of pharmacy, both ancient and modern. A quarter of modern drugs contain a primary plant constituent, and fully half are directly based on plant derivatives. Some 121 drugs on the market today are derived from plants, and 90 percent are still extracted from natural sources. Among the most famous is the Madagascar periwinkle, a pretty pink flower which is the basis for the cancer drugs vincristine and vinblastine, used to treat leukemia successfully. The list of such plant-based drugs is long and illustrious.

King has seen firsthand the lives of the 80 percent of people in lesser developed countries who rely on plants for primary medical care. In developed countries such as Germany, herbal medicine has been validated officially by the government and accepted by the medical profession. Germany's federal Commission E has amassed authoritative documentation on the traditional uses of herbs, their safety and efficacy, leading to $3 billion in annual sales of herbal medicine, half by doctor's prescriptions.

King points out that the modern drug industry is a relatively recent phenomenon which burgeoned in the 1920s based on mimicking natural products through synthetic analogs created in the laboratory. From the companies' point of view, laboratory chemistry is more predictable than plants, and, even more important, the drugs could be patented as proprietary products. Since it costs so much to develop a new drug, the companies have had little or no interest in an unpatenable herb that could be grown in someone's backyard. Moreover, screening plants for medically active compounds is a dicey process riddled with blind alleys and an incomprehensible complexity of ingredients and components. Drug companies followed a reductionist "magic-bullet" approach, rejecting the synergistic intricacy of the natural world. Plant medicine fell into disrepute as primitive and ineffectual.

According to King, the pendulum is now swinging the other way. "Twenty years ago, pharmaceutical companies were looking very hard at plants, but they went to genetic engineering and biotechnology. In fact, this direction hasn't been anywhere near as successful as they had hoped. Now there is a move back toward plants and other natural organisms."

There is evidence of this trend in the agreement made between Merck, the pharmaceutical giant, and the government of Costa Rica for "bio-prospecting" in the country's botanically rich forests.

"The tropical forests of the world," King observes, "are the most prodigious chemical factories. No matter how many rational drug designers you have sitting at high-tech computers, they're not going to approach the complexity that nature has managed to produce. Human beings who have been living in these environments for millennia are some of the wisest and most knowledgeable people about the characteristics of these plants. They are a piece of their environment, not outsiders. What we're really talking about is both biological and cultural diversity."

The irony of the situation, King points out, is that 74 percent of the main 121 natural product-derived drugs have a historical usage by native peoples for medical purposes, employing the natural products on which the drugs are based. Quinine, derived originally from Cinchona bark, is still the preferred drug for deadly malaria. Curare and other Amazonian arrow poisons are used in surgery as muscle relaxants. Pilocorpus Jaborandi from Brazil is the core of a prescription drug for glaucoma and dry mouth syndrome, and the original tip came from native peoples. The list of medical gifts to the world from indigenous people is lengthy and green.

"When I met Lisa Conte," King recalls of the entrepreneur who founded Shaman Pharmaceuticals, "and heard what she was trying to do with Shaman, I was hooked. Her vision was to make a business that would discover drugs from tropical plants, use indigenous knowledge to speed up the discovery process, validate and honor indigenous knowledge, and start a conservation organization at the same time." By placing an economic value on biodiversity, King hoped the company would add the clout of the market to the struggle to save biodiversity.

Conte, who had worked in the biotechnology venture capital community, saw an opportunity in the late 1980s to create a unique business with a social mission. King joined her wholeheartedly, and together they galvanized many of the top ethnobotanists in the country to support their plan. They successfully raised the startup capital, and commenced the ambitious task. "At first, the financial analysts viewed us as kind of wacky, and it was hard to keep away from the voodoo label. Now the same people are saying that they are not making an ethical or moral judgment, but that

107

they simply feel that this makes good business sense. It's both good ethics and good business."

The company chose to focus principally on antifungal and antiviral drugs, including a herpes drug, avoiding the politically charged arenas of cancer and AIDS, which are capital-intensive and highly competitive. Shaman dispersed King and other botanists to the tropics to begin meeting with indigenous healers to identify prime botanical prospects for drug development.

RECIPROCITY

King confronted a decidedly mixed reception among native peoples. "They've been exploited, ripped off, and abused by everybody. We offered them immediate reciprocity. We asked them what they needed, and how we could help in the short term, medium term, and long term. We involved them in the decision-making process. Our botanists paid real attention and deference to their healers. It had a powerful impact."

Historically, drug companies have come into native communities, gleaned their knowledge, and taken away the prize without compensation or acknowledgment. One of the more extreme examples of this process occurred with the Cuaymi Indians of Panama in 1993. Western scientists found in their blood samples an HIV-resistant genetic trait, which the U.S. government then tried to patent without even informing the people. The Indians were outraged when they discovered the incident, and the U.S. government later retracted the patent claim. According to King, the Indians said that it was not that they would object to helping find an AIDS treatment. They just wanted to be part of the process from the outset and have some control, along with some of the prospective benefits.

A similar incident also occurred in 1993, when the W. R. Grace Corporation patented a derivative of the neem tree of India as a natural pesticide. The neem, an Indian national treasure with a folk usage dating back thousands of years, is regarded by Indian people as their cultural property. The people of India have charged W. R. Grace with "piracy" because the company has offered no compensation or acknowledgment to the native source on its "discovery."

On behalf of Shaman, King has helped initiate a policy of corporate reciprocity for intellectual property rights based on requests by the people themselves. The company has committed to provide a portion of profits of any and all products to all the communities and countries in which it has worked. Its policies and activities were initiated in 1990, and now offer a contractual model for how to deal equitably with indigenous peoples and their intellectual or cultural property. At the outset, Shaman established an independent nonprofit arm, the Healing Forest Conservancy, to address such issues of compensation and reciprocity. The conservancy manages a database on countries and cultures in which Shaman is working and administers grants.

Because the drug development process is protracted and the needs of the people are immediate, Shaman adopted a policy of instituting benefits in the short term as well. In a Quechua Indian community in Ecuador where Shaman wanted to do plant research, the company first asked permission to search, which was granted, and then solicited the community's needs. The group requested the lengthening of an airplane runway for emergency medical evacuations, which Shaman facilitated in the spring of 1992. When Shaman began formal prospecting later in the year, the village requested one large cow to feed the community during the visit. The people were further supplied with periodic visits from a culturally sensitive physician and modest financial support for the local shaman and his female apprentice so that they would not have to work for cash at a nearby tea plantation and could continue the apprenticeship. When the community asked for standard over-the-counter medicines, Shaman complied, but also provided a *materia medica* of the local plant medicines appropriate to their health needs.

In Tanzania, the company has supported an HIV healers' network to deliver herbs to sick patients. In Thailand, the Wa people asked that their children be educated in the Thai language to deal more effectively with the government. In Guatemala, the community was enabled to electrify several buildings. In the war-torn Congo, the company funded an emergency rescue of a local herbarium filled with rare plants. In Colombia, Shaman got copies of a book on the local plants for the school, which started teaching a plant a week to the kids to reclaim their botanical history and heritage.

In all communities, Shaman is leaving all the plant data it collects with the local communities which have collaborated. "We're in about twenty-three different countries now and over a hundred cultures," King notes. "We will pay royalties on drugs successfully developed to these communities. If it's a big market with lots of money, there will be a nice cash flow going to these countries for conservation programs and health matters that they deem important. The bottom line is poverty. If you can't offer alternatives to logging and cattle ranches, you're offering your hopes and dreams without any tangible alternative. If you don't deal fairly with the people as inhabitants of the forest, you're not going to succeed in keeping biodiversity as the heart of medicine."

King finds that the exchange principle is endemic to indigenous life. "Any time that anyone went hunting, nobody ever came back and ate it alone. It was always distributed among the group, whether it was one fish or a hundred. It is a functional principle."

Despite its efforts, Shaman has come under heated criticism as a potential exploiter of native peoples. Critics charge that the proposed royalties are inconsequential at best, and doubt the recipients will ever see them. Only time will tell whether the company's efforts will indeed bring just compensation as promised.

CONSERVING FORESTS AND FOREST PEOPLES

King believes that placing an economic value on these botanical resources is helping save the land and cultures. "If you can ante up some cash right when someone's about to pay nothing for cutting down an ancient forest, a government or a speculator will take the money and leave the forest standing. That's tangible and real."

Leaving the forest standing, however, is not as easy as it sounds. When anticancer properties were discovered in the Pacific yew tree in the U.S. Northwest, the news produced a run on the forest as wildcrafters wantonly stripped the bark from the slow-growing trees, endangering the newly found natural resource. Consequently, sustainable harvesting is an essential criterion for plant-based drugs for Shaman.

Use this card as a bookmark, then tell us what you think...

Chelsea Green publishes books on a variety of subjects related to sustainable living. Return this card for a complete catalog of our books. Please indicate the subject(s) that interest you most:

___Renewable Energy ___Shelter ___Food ___Gardening ___Nature ___Environment

Other topics that interest you: _____

What publications do you read regularly? _____

In which book was this inserted? _____ Where purchased? _____

How would you describe your satisfaction with this book?

___Exceeded expectations ___Met expectations ___Could be improved ___Disappointed

Comments for the author or publisher: _____

_____

_____

Name: _____

Address: _____

E-mail: _____

☐ I'd like to receive updates and special offers from Chelsea Green.

www.chelseagreen.com

Thank you very much!

Books for Sustainable Living

CHELSEA GREEN PUBLISHING
POST OFFICE BOX 428
WHITE RIVER JUNCTION, VT
05001

Visit us at our website
www.chelseagreen.com

"Our first product could almost be considered a weed," King says of the herpes drug, "because it is so widely distributed. It is found in eleven countries and it's fast growing. We pumped in well over half a million dollars in thirteen different studies by ecologists, foresters, botanists, and anthropologists looking at how it reproduces, the ecology of the seed distribution, germination time, and species management. The broader the regional area that people are involved with, the safer it is both from a supply and an ecological point of view. Because the drug discovery process takes seven to twelve years, we have time to put in place a sustainable management system. And if the drug didn't make it, I would be very happy that we have put in a real base of knowledge in managing the species. Part of how we return benefits to the communities is the creation of new natural product supply industries on a sustainable basis."

In order to assure long-term sustainable harvesting and conservation, Shaman entered into lengthy negotiations with representatives of the Pan Amazonian Indigenous Peoples Federation of Amazonian South America, which represents 70,000 native people. Among the concerns of the group was that the people would receive the results of the research, as well as acknowledgment of indigenous knowledge as the source. The company also negotiated to pay a premium price for the plants in return for a guarantee of the quality and ecological integrity of the plants and their harvesting. The company is now pursuing similar experimental supply and purchase agreements with other groups in Peru, Colombia, Mexico, and Ecuador.

"What's really going to make this work is having many more people trained in biology sitting around making the decisions and the deals. Biology is the driving force, and acknowledging that things are so closely interwoven will prevent the system from falling apart."

King has found other considerations besides economics to be of prime importance to native peoples. "There is a spiritual dimension between plants and humans, a reverence for life and nature, particularly for the medicinal plants. You realize that this is their sacrament, a connection to a greater being. Nature is a religion, and this has been a spiritual teaching for me and many of my colleagues. Yet you find a lot of playfulness and laughter and just plain enjoyment of life."

Meanwhile, back in Shaman's laboratories, the plants are indeed panning out. According to King, researchers are finding a 50 to 70 percent correlation of activity based on the cultural knowledge of the forest peoples. "We tend to be validating their knowledge quite heavily," King says. The indigenous knowledge also gives Shaman a jump-start on the drug discovery process, saving vast sums of research dollars and speeding the ultimate availability of valuable medicines.

THE RESURGENCE OF GREEN MEDICINE

King sees a wealth of benefits emerging from the dream. Shaman's potential success will bring much greater acceptance of plants and plant-based medicines in the United States because they work. He foresees an increased respect for traditional native healers and their methods, which is already leading to greater acceptance in their own lands as well. He believes that more research funding will come from established agencies, and businesses will have a working model for increased cooperation and partnerships with native peoples and local institutions. He hopes that countries will respect their indigenous populations and collaborate more effectively with them.

In fact, all these changes are now occurring. The National Institutes of Health (NIH), the main medical research arm of the U.S. government, has reactivated its plant medicine screening program, working closely with herbal healers in Belize and other countries to identify new drugs. Bristol Meyer Squibb, a huge drug company, has made deals in Asia to work with native herbalists to look for useful medical plants. And countries are rapidly waking up to the fact that their indigenous populations and plants are national treasures worthy of respect and protection. In fact, genetic resources may well emerge as the greatest riches of all, both biologically and financially.

"My greatest hope," King concludes, "would be that we might even affect our crisis-ridden health-care system and induce another look at more moderate plant interventions, supported by insurance companies and the government. Many young physicians are eager to get training in acupuncture and Chinese herbs. Many people are going to alternative practitioners and healers. People can encourage their politicians to support the right things. On a personal level, people can grow herbs, learn about them from local people, eat healthy food, and live well—change things from the inside out."

Nevertheless, King has found that indeed it is a jungle out there, fiercer in the business world than in the natural wilderness. Apart from the severe criticism Shaman has attracted, it has also continued to struggle financially through the distended drug discovery process. The company's "burn rate" of capital is intrinsically high, as clinical trials to demonstrate drug efficacy drag on interminably, a common hazard in the pharmaceutical industry. In 1994 Eli Lilly terminated its joint venture with Shaman to look for antiviral compounds, bruising the company's budget and reputation. Shaman recently did make successful alliances with Japanese and Italian pharmaceutical concerns, but whether it can survive long enough to be successful is uncertain.

One of the shamans with whom King has worked, Ilias, graced the cover of *Time* magazine in 1992 for a cover story entitled "Lost Tribes, Lost Knowledge." King asked him how he felt about his picture on a magazine reaching millions of North American people. "He said that his father told him that we'd be coming. He predicted that some *gringos* are coming to look for medicine, asking 'Can you help us?'" King remembers that Ilias paused and nodded. "We will help," the native healer told King generously, and he has.

"Time is running out," King concludes urgently. "The bridges need to be constructed as fast as possible. If we can do that with a view of the needs of the people and the forest, the benefits will be to all people, and to the world's pharmacopeia."

The self-described "tropical tramp" is already checking his date book for the next trip to Latin America, in search of the plants and their indigenous guardians who so inspired him to seek alternative means of healing. It will be several years before Shaman Pharmaceuticals' plant-based drugs reach the market, and by then, if Steve King has his way, these precious plants and the priceless knowledge which led him to them will still be there.

CHAPTER FIVE

From Agribusiness to "Agri-culture"

ECOLOGICAL FARMING METHODS CAN PRODUCE FROM TWO TO FIFTEEN TIMES as much food as conventional chemical farming, according to biointensive farming pioneer John Jeavons. As well as producing food, these same practices "grow soil," which represents our biological capital. The 3,000-year-old Incan farming methods using water canals produce three times as many potatoes as any "modern" methods, without the use of the wheel or draft animals, and with no harm to the land.[1]

Modern global agribusiness has produced enormous quantities of food relatively cheaply, but this ostensible success has come with perilous environmental, health, social, and political costs. There is strong historical evidence that unsustainable methods of food production coupled with population pressures have been the primary causes in the unraveling of several previous civilizations.[2] Unfortunately, most of us in the industrialized world are so removed from the land and the origin of our food that we are unaware that modern agriculture is the single most environmentally destructive human activity!

Because food is the biggest business in the world, redirecting agriculture to ecological methods will produce enormous environmental improvement. It also has the potential to create many jobs, revitalize community economics, strengthen local food self-reliance, and support healthy nutrition.

There is a large body of traditional farming practices and contemporary innovation that can provide an abundance of food while also renewing the land. Ecological farming could be widely instituted based on what we already know. The models for changing our approach to farming are proven and established. Scaling them up would immediately begin to offset environmental damage and renew rural life around the world.

THE PROBLEM

The invention of farming has wrought havoc upon the environment. It has led to massive topsoil destruction, severe water pollution, the spread of infectious diseases, and overpopulation. Agriculture has also destroyed wilderness, forests, and wetlands. The food surpluses created by farming also created hierarchical power structures and widespread slavery. It was these original "bean counters" who invented writing, which was first used to record debts.

Agriculture today is the largest polluter of any industry, and the damage created by farming has intensified in this century with ever more powerful modern technologies. The Green Revolution has brought the massive use of toxic chemical fertilizers and pesticides, heavy water use, expensive heavy equipment, and "high-yielding" hybrid seeds, all of which have exhibited major flaws. Despite this supposed technological panacea, grain harvests are starting to level off and to drop, even as world population continues to swell. Ironically, farmers today lose more crops to pests than they did in the 1940s, when pesticide usage was tiny.[3] Meanwhile we are "treating our soil like dirt" and losing alarming amounts of topsoil, our very biological capital, to erosion. It is estimated that natural processes take a hundred years to replace a single inch of topsoil.

Agriculture is the largest user of petroleum of any industry. It will be among both the principal causes and casualties of climate change and global warming.[4] The effects of climate change could cripple food production, especially large monocultures, since hybrid seeds are more finicky and less adaptable to change than traditional open-pollinated and heirloom seeds. They are considered "thoroughbreds" more vulnerable to fluctuations.

Agriculture is a major cause of biodiversity loss because it destroys habitats and poisons wildlife. Less well recognized is the shocking loss of biodiversity in food crops themselves (see Chapter 2). Since the turn of the century in the U.S., up to 97 percent of the diverse strains of foods once available have disappeared.[5] An estimated 75 percent of traditional European seed stocks are believed on the edge of extinction.[6]

The corporate concentration of food-related industries is a vertically integrated package from seed to shelf known as "agribusiness." According

116

to the Federal Trade Commission, none of the four major agricultural input industries—fertilizers, chemicals, seeds, and machinery—is clearly competitive. In the United States, it is projected that by the year 2000, just 1 percent of farms will produce 50 percent of the nation's food.[7]

In the last 20 years, over 1,000 independent seed companies have been acquired by very large transnational chemical and pharmaceutical companies. These giant companies seek to control the market vertically to ensure that growers buy their patented hybrids and purchase agrochemicals from the same "company store" selling the seeds (which are bred to depend on these specific chemicals for their growth). The global loss of seed diversity is a direct result of corporate concentration and plant patenting.[8]

Even advocates of the Green Revolution acknowledge that no dramatically increased yields from technological fixes are on the horizon. They do herald the coming era of genetic engineering, but most efforts to date are directed to increasing the tolerance of plants to absorb pesticides. Biotechnology enthusiasts are blithely crossing boundaries that nature generally does not, creating "transgenic" species (breeding across species lines). The very definition of a species is that it is a closed gene pool. We have no idea what the consequences of transgenic experiments may be for health and the environment.

The emergence of "mad cow" disease might be an alarming harbinger of the potential dangers of genetic engineering.[9] In humans, the disease has been seen previously only among cannibals, and is at least partially related to feeding beef to cows, lamb to sheep, and so forth. Since genetically engineered food products now have human genes inserted into them, we will soon be eating our own species.[10] The safety of the food supply was already jeopardized by environmental pollution. Adding biological pollution to the mix poses another dilemma. Mad human disease may be next.

Meanwhile, U.S. government policies penalize farming sustainably or organically. Instead, a system of "farming the government" by giant subsidized agribusiness corporations has arisen which virtually mandates large inputs of chemical fertilizers and pesticides, as well as a narrow band of crops based on hybrid seed stocks. But at the same time that these destructive agribusiness practices are taking their toll, highly positive trends are also emerging.

117

SOLUTIONS

There are sustainable agricultural technologies that could replace conventional methods with only a brief transitional drop in yields. These methods are capable of restoring soil and improving the safety and nutritional quality of food. But decentralizing agriculture and freeing it from corporate agribusiness domination are also essential changes needed.

Sustainable or alternative agriculture is a broad tent covering a variety of practices. Overall, it is an approach designed to conserve topsoil, water, and energy, and to use far fewer synthetic fertilizers and pesticides, if any. The methods used include polycropping (mixed plantings), crop rotation, trickle irrigation, nitrogen-fixing crops, and integrated pest management (IPM) involving natural biological insect predator strategies.

Alternative systems such as permaculture and biointensive and biodynamic agriculture have shown their ability to produce equal or higher yields of more nutritious food while building soil fertility and conserving natural resources. Increasing numbers of U.S. farmers are employing one or more sustainable practices. In fact, according to many USDA studies, the most efficient farming units, from both economic and environmental perspectives, are small- to medium-sized family farms, which are enjoying a resurgence based mainly on value-added organic and specialty crops using diverse seed stocks.

Clearly a growing segment of the public is voting with its pocketbook in favor of organic foods, diverse varieties, and family farms. Demand for pesticide-free and organically grown food has boomed to an estimated 3 to 4 billion dollars worldwide. In Europe, the number of organic farms doubled between 1979 and 1991. In California, Pandol & Sons, Gallo, Fetzer, and other major grape growers have been slashing pesticide use on their crops without loss of yields and with improved quality. About 2,000 farmers markets have sprung up across the U.S. in just ten years, bringing farmers and customers together to celebrate farm-fresh food, family farms, and community. Community Supported Agriculture (CSA) is emerging across the land as groups of people link with local farmers and "subscribe" for a season of fresh food.

118

Progressive national companies such as Odwalla, Stonyfield Farms, Wild Oats, Eden Foods, and a plethora of others are starting to work directly with family farms to stabilize their supply and improve the reliability of quality. Connecting farmers directly with these alternative markets is an ongoing process that is building a positive agricultural economy outside the commodity markets.

Small groups and companies such as Seed Savers Exchange, Native Seed/SEARCH, Of The Jungle, Seeds of Change, J. L. Hudson, and Seeds Blum have been inspiring and empowering thousands of farmers and gardeners to practice "backyard biodiversity." To diversify our narrow food choices, they are reintroducing hundreds of nearly lost species and highly nutritious or medically valuable crops from other cultures. Meanwhile, in a different form of resistance (see Chapter 2), half a million farmers in India militantly demonstrated for their "sovereignty over seeds" and against GATT seed-patenting provisions. This antipatenting movement supporting agricultural biodiversity and sustainable family farming is spreading rapidly throughout the lesser developed countries, challenging the very basis of agribusiness.

WHAT YOU CAN DO

Eating is something that most of us are fortunate enough to be able to do each day, and you can make a real difference in restoring agriculture through your daily choices. Select organic options whenever possible. Try fruits, vegetables, and grains that are unfamiliar to you and learn about them. Support the stores and markets that offer these alternatives.

Find out if people in your neighborhood have a Community Supported Agriculture project and join, or work with others to start one. Connecting with the farmer who grows your food is a deeply satisfying experience, and most CSAs permit you to go and pick your food if you want to. Visit your local farmers' market and enjoy the colors and scents and people. Food fresh from the source handed to you by a grower with dirt under his or her nails is a real celebration. Plant a garden or a window box.

You can also subsidize companies, stores, and restaurants that support organic farming and community-based farms. If they don't, ask why not. If you're so moved, you can also start a farm or greenhouse like those de-scribed in the following pages. It's not an easy life, but if it's right for you, it's very fulfilling. Restoring this most basic link in our relationship with the land is something that brings great pleasure to millions of people. Whatever you do, get involved.

All over the world there are sophisticated traditional farmers and gar-deners who have the know-how to implement a sustainable agricultural system. There is a growing public movement for a safe food supply and a decentralized system of ecologically sound food production based on fam-ily farms. The growing size of the markets for organic and specialty foods is now allowing these movements to become an industry with political im-pact. As awareness heightens about the negative environmental impact of agribusiness farming practices, its methods will no longer be considered ef-ficient, "cheap," or tolerable.

In this chapter, bioneer farmer/gardeners Fred Kirshenmann and Anna Edey provide striking models of new ways we can feed ourselves abundantly and support farmers while minimizing harm to the environment. These el-egant solutions point the way to a positive farming future that restores our land, food, and communities. As these bioneers demonstrate, we can feed ourselves well while living lightly on the Earth.

Fred Kirshenmann
and the Soul of the Soil

*Fred Kirshenmann is a midwestern family farmer who has made an
ecologically and financially successful transition to organic biodynamic
farming. He has also pointed the way to helping organize farmers into viable
economic entities by bringing them closer together with consumers with
value-added products. He is affecting public policy to help spread organic
farming and decentralize food production into regional areas called foodsheds.*

IT IS A FRIDAY AFTERNOON IN JANUARY AND FARMERS HAVE BEEN DRIVING
hundreds of miles to Aberdeen, South Dakota. Converging on the Northern Plains Sustainable Agriculture Society annual conference, they have
crossed the Upper Midwest's vast frozen expanses, glistening like frosted
glass puddled with serpentine lakes amid the unmistakably human grid of
right-angled farms. These organic farmers are gathering for an advance
meeting on an issue of dire import: whether to pool their collective resources into a cooperative. The coop will give them power as a group to
reach markets directly and command fairer prices. It is the culmination of
a three-year effort, funded by grants from North Dakota state government
and the farmers themselves.

The group is varied, with pony tails and crew cuts, wearing flannel and
polyester, overalls and suspenders, sneakers and cowboy boots. Now they
must see if they can bridge their diversity to express a unified voice, a modern David struggling in a Goliath marketplace.

A farmer stands by a blackboard writing down numbers as he exhorts
the room. To be financially viable as a marketing entity, the coop needs a
minimum guarantee of twenty thousand acres. For many years now, there
has been a war of attrition as farms held in families over three and four
generations have failed financially, only to be gobbled up on the auction
block by giant landholders who control modern agribusiness. These proud
farmers are desperately straining to salvage not only their own survival,

but also a noble lineage of self-reliance and caring for the land. The odds are against them.

Where once 6 million farmers worked the land in the 1950s, fewer than 2 million farm today. Federal farm policy sets commodity prices below production costs through price supports that favor giant grain traders. A mere four giant corporations control almost half of agricultural commodities. Farmers, known for their insistent self-sufficiency, are today marginalized as contract laborers for the company store, and many here today are fearful of losing even that crumbling toehold.

Terry, a farmer of Norwegian heritage, scratches digits on a blackboard, but the numbers are coming up short. The farmers are scared. When they launched the idea for the coop three years ago, commodity prices were down. This year they are back up, and people are anxious about putting their chips into an unknown entity outside their control. Terry implores the group to remember that community is our strength. He says that commodity traders have already shown up at the conference, and are approaching the largest farmers, playing the old game of divide and conquer. "Divided," he says, "they pick us off, farm by farm. Together we will prevail."

MAVERICKS IN THE HEART OF AGRIBUSINESS

Fred Kirshenmann, a founder and board member of the society, watches intently, his chin cupped in his thick-boned farmer's hand. His red and gold hair stands out in the crowd among the many blondes, grandchildren of European immigrants from Scandinavia and Germany who arrived on the Great Plains a century ago or more. He is among the largest organic farmers in the Midwest, and he knows that the risk they are today considering is less than the biggest risk they all took years ago: going organic.

These farmers are mavericks in the heart of agribusiness. At this time of year, the ads on TV for chemicals are as thick as Johnson grass. These are not ads seen by anyone in New York or San Francisco. They are paeans to chemical agriculture. Ads with children frolicking through tall green grasses, happy grandparents, majestic tractors, and elevating music end with a soft sell for Atrazine, "the only chemical a farmer needs." Atrazine is one of the deadliest carcinogens on the market, now found in drinking water across the Great Plains, saturating the precious Ogallala aquifer, whose water is being drained for massive irrigation at a rate 160 percent above its ability to replenish itself. In former years, ads were more blunt. One showed two worms wiggling toward each other through the soil as a huge bag of poison explodes on them from above. If the ads don't convince these skeptical folk, the really hard sell is that farmers can't get a bank loan or qualify for various federal price-support programs unless they contractually agree to use certain chemicals.

Fred Kirshenmann was born in 1935 in North Dakota at the height of the notorious Dust Bowl, when you couldn't see from the house to the fence post for the clouds of dirt fogging the air like some malefic dry rain. European settlers fleeing poverty and oppression staked out land formerly occupied by Native Americans, and in the span of a generation, the sodbusters broke up the shallow topsoil. When the drought came, it was topsoil that was blowing in the wind. Fred's father Theodore swore his boy would have a better life—off the farm.

But Fred grew up with the love for a land that his father imparted. The family had come from Europe, Germans who settled in Russia after the Six Years' War when Catherine the Great invited Germans to come settle and farm in Russia. She neglected to tell the uprooted peasants that their new home was a nest of brigands and thieves so thick that even the police dared not venture there. Fred recalls his grandfather's tales of going out into the fields with the reins of oxen in one hand and weapons of self-defense in the other.

As the Bolshevik revolution mounted and the czar conscripted all available citizens, the insular German expatriate community resisted, choosing jail or immigration. It tenaciously held on to its identity, as reformed Protestants among Lutherans. When Fred's grandparents migrated to the United States a church network guided them to Nebraska to homestead land. Reliance, Nebraska, had beautiful, rich farmland, and Fred's father remembers leaving it at age six. It was level ground with no rocks, and the topsoil was fertile and deep. But the family wanted community and migrated further north to join the German-Russian town of Streeter, North Dakota. Fred recalls with bemusement, "My father says that because his father thought we were going to marry heathen Lutherans, he brought us up here to these rocks!"

Fred's father and mother started farming in 1930 at the onset of the Dust Bowl. "It made a lasting impression on my father, and he learned first-hand what could happen if you didn't take care of the land. I remember throughout my childhood growing up on the farm how my father used to get almost hysterical when there was any wind erosion, and he would do anything he could to prevent it. It wasn't till later that I really understood why he got so excited about a little dust blowing in the fields. I grew up with the notion that it was really, really, really important to take care of the land."

The land is the great northern prairie, rolling hills, and grasslands dotted with potholes filled with the precious water that attracts flocks of geese and ducks, even seagulls and pelicans. The prairie diversity, though much diminished today, is still rampant with a variety of grasses. Deer, foxes, and coyotes prowl in abundance where once the buffalo roamed.

"WILL THESE FERTILIZERS HURT THE LAND?"

Theodore considered himself progressive, and the farm was diversified with grain crops and livestock, as it is today. He was always investigating the newest technologies. He often said, "If you don't progress, then you don't have a place on this planet. That's what we're here for is to progress." As a result, Fred's father was the first farmer in the township to start using fertilizers and pesticides when they became available in the late 1940s.

"I remember as a youngster," Fred says gently, "that he went to the county agent and to several progressive farmers to talk about whether these fertilizers were going to hurt the land. That was his main concern. In fact, the county agent said it would be good for your land. He went ahead on that recommendation."

Although Fred knew he was destined to leave, he loved the farm. "I started driving a tractor when I was seven and haven't been off it since. I had a deep love, especially in the spring, for watching the plants come up through the ground and seeing fresh soil. Those were very visceral experiences for me. But my parents were always very insistent that my sister and I get all the education we could because they were deprived of that. In my grandfather's family, if you had enough education to calculate your grain receipts, as a boy, that was enough. As a girl, if you had enough education to keep the household records, that was enough. There was never any question that I would go to college."

Fred earned a B.A. in philosophy and religion, and a Ph.D. in historical theology. He went on to teach at the college level and returned to the farm each summer. During his teaching days at Yankton College in South Dakota, he learned about organic agriculture from a pupil, David Vetter, a graduate student in soil science. With great excitement, Vetter introduced Fred to *Agriculture*, a book containing a series of seminal lectures by Austrian scholar Rudolf Steiner.[11]

During his tenure at Yankton College, a church-related school where he taught philosophy and religion, Kirshenmann met Janet Robinson, whom he would later marry, and they moved to Massachusetts, where he served as dean of students at Curry College. They spent summers back at the farm, helping the aging Theodore with the harvest. In 1977, a telephone called changed all that.

Theodore had had a mild heart attack. "It was a signal for him that he needed to change his lifestyle and get out from the pressure. He announced that he was going to try to find somebody to manage the farm." Disenchanted with urban life, Fred and Janet wondered if this might be what they were looking for. "I called my father and said that if the job was still open, and if he would let us convert the farm to an organic farm, we'd be interested in coming out and managing it."

GOING ORGANIC BY TRIAL AND ERROR

Kirshenmann's studies of organic farming had convinced him of its importance and viability. He told his father that the comparison data he had seen showed how organic farming improved soil quality. Theodore acknowledged the breakdown of soil structure on the farm from chemicals, but the elder Kirshenmann was skeptical. "He was very quick to realize that it would require a change in the whole system of farming," Fred explains, "that it wasn't just a matter of switching one set of technologies for another. He was sixty-eight years old, and he just said, 'You know—this is for somebody else. I'm too old. If that's really what you want to do, I would be delighted.' "

Although Kirshenmann knew the 2,300-acre farm well, he knew little about making a transition to organic, and even less about marketing. He went to the county extension agent seeking information. "At that time county agents were not interested in organic farming at all. His direct comment was, 'Organic farming—yeah—I guess there's a few people that do raise some food for health food stores.' That was the extent of his knowledge or interest. We learned by trial and error."

Kirshenmann boldly converted half the farm to organic the first year, and the universe smiled. The rains came when they were needed, and the frost stayed away till late into the fall. He brought in a large healthy crop, and decided to go all out the next year, completely converting the farm to organic. This time everything went astray. Early-growing weeds overtook the crops, and yields dropped by a breathtaking two-thirds. Fortunately, the farm had no debt and even had some cash reserves to tide them over.

"Out here neighbors seldom tell you what they think," Kirshenmann muses kindly. "They always tell other neighbors, who then tell you. Apparently the talk around the neighborhood was that we were crazy, we were going to lose our shirts. You could never farm without chemicals. How did this kid get so corrupted on the East Coast? Your mistakes are always out there for everybody to see, and that pretty much confirmed their initial judgment that this would be a disaster."

Kirshenmann began to identify the particulars of sustainable organic farming. He successfully substituted nitrogen-fixing legume cover crops for synthetic fertilizers. He rotated crops among fields to discourage the bugs that thrive on the same moncultural crop in the same field year after year. He planted clover as a green manure to turn back into the soil for nutrients and tilth. He also used the clover as mulch, blanketing the fields with vegetative cover to preserve moisture and nutrients. His herd of cows provided abundant manure for fertilizer. "We might run out of oil, phosphorous, or potash," he comments wryly, "but we'll never run out of manure." Over four years, he achieved a balance of soil fertility and productive yields and expanded to 3,100 acres.

"Then as we started to work our way out of the problems, our successes were also very obvious. I remember a couple of neighbors coming over and looking at the wheat, and they'd say, 'Well, what kind of variety was that you planted in your field?' The implication was obviously that it had to be the variety of grain, not the management, that caused it to be successful. When they found out we were using the same wheat varieties they were, they would just look puzzled and walk away."

The word on the street was that Kirshenmann's organic farming was interesting, but could never work unless everybody did it. The assumption was that without fertilizers and pesticides, the farmer would have to take a drop in yields, which would be all right if everyone had lesser yields, causing the price to rise. But the bigger puzzle was that the Kirshenmann family farm's yields dropped only temporarily, then rose above conventional ones. Farmers still balked because adopting organic methods would disqualify them from the federal programs upon which they were dependent. Many farms were also heavily leveraged with debt, and just one bad season could send them to the auction block.

Kirshenmann was fortunate to have a positive relationship with a community bank and a banker who displayed a genuine interest in organic farming. With a $50,000 credit line, Fred built the farm successfully and has needed no outside funding since 1980. He continued to visit the banker, who became a firm advocate of organics—or "farming without pesticides," as he put it. At that time, banks were encouraging farmers to borrow money freely, and as petroleum prices skyrocketed and commodity prices fell, more and more farmers were feeling the crunch. The banker now encouraged farmers in difficulty to look at Kirshenmann's model of reducing input costs dramatically without jeopardizing their yields.

Kirshenmann started to gain a new understanding of farm economics. The only ways for a farmer to recover value are to reduce input costs of expensive petrochemical fertilizers, or to add value to the product in the marketplace.

By using organic methods, Kirshenmann created a more flexible biological system less dependent on outside economic forces. But the process also entailed a different relationship with farming. "I wouldn't say so much that nature is your friend, because I quarrel with nature almost every day. But nature is a partner. You do a dance with nature, and you always have to be tuned in to what kinds of steps she's going to take, and then figure out how your steps are going to fit so you can do the dance together. When we plant a legume cover crop, we can't follow it with a cash crop in the same year, but it provides a fertility base for the next four or five crops. My sweet clover costs aren't going to go up or down that much either. If the seed gets too costly, I can harvest and grow my own seed. The system is less brittle, less vulnerable, more flexible. I'm never going to get bored with this job, because there's always something new that comes along."

THE BIODYNAMIC MARKET

Finding markets for organic products, however, was a more perplexing matter. Running a large farm is work enough in itself, and farmers generally don't have the energy or disposition for marketing. Kirshenmann tracked down East Coast-based Mercantile Food Company, a distributor of organic foods. He worked out an arrangement for the company to have first dibs on all his crops as long as it would try to sell all of them, at a hefty 15 percent premium because of their exceptional quality. Market considerations now entered the equation, and as Kirshenmann discussed prices with Mercantile's president, the trader pulled a well-worn book off the shelf. It was the same Rudolf Steiner's *Agriculture* he had seen years ago. The distributor told him that switching to biodynamic growing was a market advantage. "I took the book home and read it again. As a philosopher, I was attracted to it, and I wanted to know if it could work."

Rudolf Steiner was an Austrian clairvoyant and scholarly authority on the work of Goethe, whose Romantic vision was also of a partnership with nature. Steiner had grown up among peasants and believed that the greatest wisdom "grew out of a peasant's skull." He retained an intimate connection to nature, and founded a spiritual school called Anthroposophy, as well as the elaborate Waldorf educational system for children. In 1922 he was approached by farmers deeply worried about the decline of soil fertility and seed stocks. The new chemical and mineral fertilizers seemed to be damaging the health of farmland, and they asked Steiner to advise them. In 1924 he delivered a series of lectures which formed the basis for what became known as biodynamic farming.

Steiner placed profound importance on the soil, but extended farming into the realm of what he called cosmic influences, the invisible spiritual life forces of the universe. His recommendations included many practices that seem outlandish and bizarre, though empirical practice has shown them to work. Steiner emphasized the making of compost, decomposed organic materials for soil fertility, and described a process for making "spiritual manure." He instructed the farmers to stuff a cow horn with fresh cow manure, and bury it in a field over the winter. Upon digging it up, he showed the manure to be a dry, odorless powder, which he emptied into a barrel filled with rainwater and began to stir vigorously, first in one direction, then sharply in the opposite. The sudden reversal created a spiral vortex like water going down a drain, which Steiner said captured cosmic forces and "potentized" the manure tea with etheric energy. He showed them how to spray the fertilizer in tiny amounts over large fields.

Steiner also gave precise instructions for the similar brewing of several fertilizing preparations, known today as the mysterious biodynamic "preps," based on herbs such as valerian, horsetail, chamomile, and nettles, as well as ground crystal quartz, each with a very specific application. He showed how to plant by the stars to maximize the benefits of cosmic forces.

Steiner also emphasized the "individuality" of each farm, laying out a program to achieve a correct balance of plants and animals, and an integration with each farm's unique ecosystem. He believed that a properly run, healthy farm needed no outside inputs at all.

Recent studies reported in the *Wall Street Journal* have corroborated the exceptional productivity and flavor of biodynamically grown foods. Large markets exist in Europe today for biodynamic products, which command a hefty price in the marketplace. So when Kirshenmann successfully adopted the Steinerian system, Mercantile was able to sell his biodynamic crops at a premium in Europe.

Despite his personal success, Kirshenmann is cautionary about a societal transition to organic growing. Although reducing fertilizer and pesticide use is imperative, an overnight switch would be disastrous because many farmers simply would not have the knowledge and tools to do so effectively. In addition, the development of organic markets is a longer term process. But there is more at stake, he believes. "We're talking about a systems change in agriculture that will not only change what we do on the farm, but also the way we eat."

FOODSHEDS

Kirshenmann suggests a new concept of food and agriculture in the shape of "foodsheds," regionalized, customized growing with a diversified set of crops oriented to local conditions and communities. He sees a decentralized agriculture reverting back to many smaller family farms, which are responsive to specific foods people want. "A foodshed in North Dakota is going to look different from one in New England. Instead of raising wheat and barley in North Dakota, we're going to raise some root vegetables and other basic foods people eat. It would have value added to it by definition because you remove a whole bunch of links in the chain between the farmer and the food customer."

With a foundation grant, Kirshenmann and others recently pioneered a new foodshed label in the Northern Plains. It signifies that 51 percent of the ingredients in a product were grown in the five-state region. It is a tangible step in decentralizing agriculture and supporting local growers.

Kirshenmann points to the encouraging fact that 2,000 farmers markets have sprouted around the nation in just the last ten years, clearly demonstrating the public's desire for fresh, often organic foods directly from the farm. He also notes the growth of Community Supported Agriculture (CSAs), known as subscription farming, where 20 or so customers band together and hire a local farmer to grow produce for them over a season. The partnership ties people directly back to the literal roots of their food, while supporting a farmer against the vagaries of the marketplace. "People need to be involved," Kirshenmann observes, "in what it really means to grow food. If you don't have a garden, use your patio or grow a tomato in a pot. If there's not organic food at your supermarket, ask the manager why not. The more that people start doing that, the more they're going to change the food system and the food that's available to them.

"Food is not like any other commodity," Kirshenmann continues passionately. "Food is a community creature. Food has always been at the center of community celebrations—a wedding, a birthday. So the industrial giants who want to completely commodify our food and reduce it to roughage for profit are bucking against a very powerful cultural phenomenon—hospitality. But true hospitality emerges when we each bring something to the table."

Kirshenmann himself helped found Farm Verified Organic, a certification agency that works with organic farmers all over the world. He also emerged as an articulate spokesperson in national public policy circles, and has participated in numerous struggles over federal farm policy and the Farm Bill. He points out that the nation of Denmark gives farmers a fiscal safety net while they make an organic transition. He has helped bring the family farmer's story to environmental groups, chefs, and church networks.

THE MINISTRY OF THE SOIL

But on Saturday night at the Northern Plains conference, he is just Fred as he introduces a special event, the screening of the rough cut of a film called *Feast of Food* featuring the story of his farm. The hall is filled with farmers and neighbors, and the film transfixes their attention. It is their life on screen. The movie intimately portrays the deeply human experience of farming in partnership with the land. Kirshenmann is seen grasping a fistful of earth from a chemically farmed field nearby. It is compacted and desiccated, and it dissipates in the soft summer breeze. He then takes the organically farmed earth from his own fields. It is dark, soft, crumbly, rich, moist. The camera does not get away without Theodore telling how Fred switched to organic and why you should, too. The proud father is today his biggest advocate. The circle is unbroken.

The filmmaker, Miranda Smith, has intercut Kirshenmann's story with her own. Her father was a dedicated farmer who died when she was only a year old. Through poignant grainy old home movies, she traces his life. He went whole hog for the USDA chemical program, spraying prodigious clouds of DDT and other poisonous inputs on his Florida orchard. He was so enthusiastic that the USDA made him a special testing agent. He mixed his own formulations by hand, without even nominal protection. A robust, happy man with a loving family, he died at the age of 40 of massive organ failure. At the time, no one could say why, but the circumstantial evidence is compelling. The farmers in the audience move a little closer to their wives and children.

The farmers are gathering again for a final meeting of the weekend on the coop formation. They are still short a few thousand acres, and emotions are running high among these people who are usually silent about their pain. This year maybe commodity prices are reasonable, warns a coop advocate, but what about next year and the year after? Agri-corporate vertical integration is rising like a thermometer in a heat wave, as giant companies devour enterprises in a feeding frenzy. They're going to keep squeezing us, and one by one they pick us off. He stares down the group. Are we a community or not?

A grizzled farmer in a worn cowboy shirt speaks earnestly. "I've made some bad investments in my time. I don't know if this is good or bad. I don't know if it will work. But this has much more chance of doing good. I've lost that much in a load of buckwheat. I'm committing." Applause erupts, but the acreage is still shy. Finally, it is clear that there will be no resolution today. The mood is despairing as people leave the room.

It's Sunday morning and Kirshenmann is in a conference pulpit of his own design. Talking about mythology and agriculture, he uses scholarly, theological terms like exegesis and eisegesis, revealing his seminary training. He relates the story of the Garden of Eden in the way it is traditionally understood. God gives Adam and Eve a paradise to live in forever, but forbids them to eat the fruit of the Tree of Knowledge. The serpent insinuates that the Lord is just trying to prevent them from eating the fruit that will give them the same power as God. Eve yields to temptation, and gets Adam to eat the fruit, too. Suddenly they both sense that something is wrong, and in shame they cover their naked bodies. The Lord comes and sees. Adam blames Eve, Eve blames the serpent, and the Lord curses them all forever.

Kirshenmann reads a letter to the editor from an agricultural magazine responding to an article about organics. The letter says that of course we have to use chemicals in farming because the land is cursed and we live in a fallen world. Kirshenmann traces the letter's cosmology back to the original environmental sin of Adam and Eve.

What about another way of telling the story, he queries. God created the garden and it was good. It was for the benefit of everyone and everything, and we are now living in the Garden. God cautioned not to live in ways that separated people as "better" than others. Perhaps the original sin of the garden is arrogance, Kirshenmann conjectures from his ministry of the soil. Perhaps we must see ourselves as one with all creation.

The conference is officially over, and farmers are heading off down the endless highway to their far-flung spreads. But a small core of people has remained to participate in an ecumenical closing service. There is anguish in the room. People write down anonymous prayers on index cards and others read them aloud. Three are prayers for farmers keeping the land, for forgiveness and compassion.

David, a lean 40-year-old farmer whose heart is broken over the languishing coop, stands to tell a story from scripture. Each day farmers come to the marketplace in Jerusalem to hire day workers for their fields. They always pick the young and strong, who must work from six in the morning till six in the evening. Left behind are the weak, the old, the halt, and the lame, with no other hope for work to feed their hungry families. One day like any other day, the farmers chose the strong and left the weak. One man, old and sick, did not leave, but waited and waited as the sun passed overhead, even as all the others gave up and left. He was desperate for money to feed his sick child. With only an hour left till work would stop, a farmer returned and hired him.

When the work was done at six o'clock, the farmer paid all the workers the same amount, a single silver ducat. The other workers protested, asking why the farmer should pay the weak laborer the same for only one hour of work. The farmer replied that it was his money to do with as he chose, and he chose to be generous. Still, the workers were angry, and they went to Jesus to ask his counsel. Jesus said enigmatically, "The first shall be last, and the last shall be first." David concludes that perhaps this is a lesson in the compassion of community, as only a farmer could know it. We have to stick together, he says, struggling against his own shaken faith. The coop *will* happen.

The transition to organic farming is a long row to hoe, Fred Kirshenmann and these farmers know.

Anna Edey's Chicken Breath

Anna Edey has created solar greenhouses capable of growing intensive quantities of fresh organic foods on a community basis. Her methods will support local entrepreneurs to produce regional and local organic food supplies, and can be retrofitted to many existing buildings.

"ONCE UPON A TIME, ABOUT SIXTEEN YEARS AGO, I WAS LIVING IN A CHARMING little cottage in the woods," begins Anna Edey with the singsong lilt of her Swedish ancestry, speaking of her beloved Martha's Vineyard island. "Here I stumbled on my first clean, green discovery," she continues, dropping her voice and leaning close over the homey kitchen table amid the enchanting fragrances wafting from the adjacent greenhouse.

An avid reader, she was reeling from the environmental horror stories of Rachel Carson and Helen Cauldicott, trying hard to envision a positive alternative. "This cottage had no indoor toilet, but it did have a comfortable setup a little distance away under the canopy of tall hickory trees. This is where we went to pee in the morning, even in the winter. Odd though it may seem, it was invariably a wonderful experience. For peeing, we used a special pot, and in order not to kill the plants or cause odors, we diluted it and tossed it here and there around the cottage.

"But I was in no way prepared for what happened that first spring," Edey says, pulling her soft graying hair back from her youthful face, her eyes wide. "In a wreath encircling the cottage, the wildflowers grew to be considerably larger and more vibrant than those farther away. Why? Was there something beneficial in the concrete used for the stonework of the cottage? Or, could it possibly be the diluted urine?"

Edey pulls out a photo of herself standing next to an Indian pokeberry bush. Pokeberry would normally grow about four feet high, yet in the picture it reaches a good three feet over the head of this tall woman. She pulls out another picture, a rhubarb leaf so big she can hide coyly behind it. For a couple of years, she used nothing but urine fertilizer for her garden, and called the practice "Urine Charge" and "Pees on Earth."

134

Edey's interest in gardening and the magic of nature blossomed. With excitement, she moved into another house nearby on the New England island and began to convert it to a solar home, using natural cycles to guide her material life. Then catastrophe struck. The airtight woodstove exploded, and the house burned down. Devastated, Edey remembers thinking it was a "slap in the face from God saying, 'You ridiculous little thing. You think you can do something to help save life on Earth?' Then I realized it was just a cleaning of the slate, which said, 'Now you'll be able to do it right instead of fiddling around with a leaky old house.'"

SOLAR DREAMING

For a year and a half, Edey scoured the world for information on sustainable design. She wanted an ecological house that not only used as little petrochemical heating fuel as possible, and caused the least wastewater pollution, but also one that would permit her to live in a garden year round and produce most of her own organic food. She visited the most advanced solar and sustainable living models she could find, including the famous New Alchemy Institute in nearby Woods Hole, where John Todd and associates were developing prototypes of such systems.

"The house ended up being much more successful than my highest dreams," Edey says with a beam, plucking a tomato off a suspended lattice in the greenhouse cum kitchen. She had been told that one couldn't grow any significant amount of food in a home greenhouse, that the temperature was always too hot or too cold unless house and greenhouse were separated by a wall. She was told that the insect problems in interior greenhouses were insurmountable, that you have to spray the plants with poisons to control the pests. She was told that maybe she could grow a few herbs, a few specialties, but certainly not tomatoes. "I ended up with arbors of tomatoes here, going all the way into the living room, going to the far end of this long skylight along the whole south side of the roof here. They were blooming and fruiting all year long." She is justifiably proud of her composting toilet and graywater system from her sink and bathtub which water and fertilize nonedible plants and trees outside. She does not, however, use human urine or waste for her food plants.

135

Edey actually started growing about 50 kinds of vegetables—celery, broccoli, cabbages, potatoes, onions and many different greens and herbs. "The tomatoes were what moved people to tears, though," Edey recalls. "People would come in here on a cold, sunny winter day and see my cold woodstove and realize that a house like this makes it possible to have a garden inside year-round. I spent less than a hundred a year on heating, using a cord of wood and old newspaper to warm all three thousand square feet. People got this incredible sense of relaxation and hope and faith, their eyes glowing, so grateful for being personally able to experience the proof that you can live in the Garden of Eden, and it's within reach of anybody."

BLISS IN THE HEART, PEACE IN THE SOUL, TOMATOES EVERYWHERE

Edey, who has lived in the house for fourteen years, was not satisfied, however. "Winter outside, summer inside," Edey muses. "In the heart of my indoor garden, there is a fifty-dollar bathtub, deep to cover the knees and go up to the nose. Now there is an opportunity for bliss, to begin to feel like maybe the world can make it. It costs nothing to reach out and pick fresh greens from the salad bed right next to it. I can reach up and pick a tomato, and most of all, I can relieve stress from my mind while I'm soaking in this water that's got leaves from marvelous fragrant rose and peppermint geranium. Now that is bliss in the heart and peace in the soul."

New England relies almost 100 percent on food from far away in the middle of the winter. Lying there in the bathtub, Edey was thinking, "I want to do more. I want to show that we could grow food without any heating fuel and any pesticides all through the winter."

Apart from not wanting the frequent visitors traipsing through her home, Edey also wanted to demonstrate the viability of a community greenhouse for her island home of Martha's Vineyard. The Vineyard is a small community of 14,000 hard-core New Englanders which swells to 80,000 in the fabled summer weather of warm Gulf Stream waters and cool sailor's breezes. But for most of the year, it is a dank, gray, sunless dab of land in the frigid swirl of the North Atlantic. Except for some patchy summer farming, the island is dependent on food from outside, mostly from

California and Mexico. Edey felt it was time to prove that even the Vineyard could be ecologically self-sufficient for most of its food and energy.

When infighting shattered the prospect of a community greenhouse, Edey serendipitously found an alternate path. A relative sent a copy of *Science* magazine with the astonishing story of an Oregon greenhouse grower who had faced bankruptcy because of his high energy bills. He removed his oil furnaces and substituted 450 rabbits. Their high body heat immediately generated 750 dollars a month in energy savings. "This was just like an explosion inside my brain and heart," Edey recalls, rolling her brown eyes heavenward. "I realized, of course, warm-blooded animals! The old people knew that, from the cave days till fairly recently. In Switzerland, Sweden, and elsewhere, they had the animals below or adjacent to the housing, common sense that people have not been practicing anymore. I spent the next twenty-four hours designing the greenhouse that I wanted to create."

AN APOCALYPTIC INITIATION

Edey devised many innovative designs, including quadruple window glazing and walls of water to hold the sun's heat, to which she gleefully planned to add her warm-blooded fuel: chickens and rabbits. After nine months of planning, she obtained a $50,000 bank credit line secured by her land. The 3M company agreed to donate their latest, best greenhouse glazing, and she sent out flyers through her network, seeking labor in exchange for room and board. "It ended up being a most extraordinary community experience. The greenhouse was built while my house was full of people. We started at the end of July, and by Thanksgiving it was mostly closed in and insulated. Then winter came like whammo just before Christmas, like a freight train from Canada with sixty mile an hour winds carrying nothing but ice cubes. The temperature plummeted to four below zero."

Edey's car died under two feet of snow. At four degrees below zero with 60-mile-an-hour winds, the greenhouse got an apocalyptic initiation. "It was such killer conditions that, walking from the house to the greenhouse, you were risking your life. I did that in the middle of the night several times. I'd wake up from the roaring wind and ice sliding off the roof or crashing against my window at four in the morning. I'd lie there and think,

'O my God! The greenhouse is surely going to freeze tonight.' Rather than lie there and worry about it, I put all my layers on over my pajamas and wrenched open the door, closed it behind me, and walked through the field with the surface occasionally supporting me—the surface being ice—and then sometimes just plunging through to above my knees. A trip that usually takes about two or three minutes across the field took me more than twenty minutes. The clouds were racing across the bright moon and the whole landscape was glistening with black shadows and brilliant white snow from the moonlight, with snow flying across horizontally from snow drifts. And me saying, 'I could die here. If I take a wrong step and break my ankle, break my knee, I could freeze to death. Dear God! I can feel you in here. I can see you everywhere. Please help me.'"

When Edey forced the greenhouse door shut behind her against the howling Arctic wind, she expected the temperature to be subfreezing. Instead, she warmed to a cozy 50 degrees where the 30 rabbits huddled snugly like a giant furry comforter. In the adjacent plant room, the air was a surprisingly temperate 45 degrees. The tomatoes were growing happily, and the whole room was bathed in the fragrances of nasturtiums, flowering ginger, and the living soil in which they grew. As she reached the chicken room, the 100 birds had elevated the temperature to a balmy 72 degrees. Their body heat was permeating the water wall to warm the greenhouse, which did not even need any backup heating. In fact, the system had only the most elementary form of backup: a large 600-gallon tub of water with a place to build a wood fire under it. Edey used it eight times that winter, burning wood scraps, and has never used it since.

Edey's greenhouse grew 100 tomato plants to 14 feet and many greens and herbs, servicing the happy islanders abundantly. With perfect timing, her crop of organic tomatoes came out just at the time of the big scare around the pesticide EDB, when supermarkets were ordered to remove or list the foods containing the toxic agricultural chemical. Suddenly, organic foods got an unexpected boost.

ORGANIC VERTICAL INTEGRATION

Then Edey began in earnest to analyze the financial viability of her system. Salad greens provided the best income. She developed an innovative system in the greenhouse to maximize the use of space by using grow-tubes, soil-filled plastic plumbing pipe, hanging from the rafters. She packed in nine tiers of tubes in the A-frame greenhouse, which she likes to call "true vertical integration," in addition to beautiful raised beds on the ground level. She fertilized them generously with compost she made from the animal bedding. She placed the plants very close together. She soon found the wide variety of crops excelling—over 100 kinds of greens, herbs, and flowers. One collard plant leaped to four feet wide and tall—a single leaf could feed a family. It finally had to be cut down after three years—with a chain saw. Apart from Edey's obvious green thumb, the plants were also benefiting from the generous carbon dioxide coming from the breath of chickens and rabbits.

Edey began to focus on specialty greens, about 50 varieties she liked. "Salad greens are an incredibly important food," she says forcefully. "It's probably one of the most health-promoting foods we can eat." Her colorful bouquets of organic salad greens were so mouthwatering that she quickly found restaurants willing to pay a premium 16 dollars a pound. She picked and processed them with great care, adding herbs and edible flowers, shipping ready-to-use boxes to the delighted chefs. She devised a system of picking only the leaves rather than whole heads at a time, thus raising productivity per square foot. When California competition followed suit, the price dropped to about 12 dollars a pound, still a healthy price.

Edey proved that her 3,000-square-foot greenhouse, a mere 100 feet long on a thirteenth of an acre, combined with a one-sixth acre garden outside, was able to bring in substantial income. "You can pull in more than one hundred thousand dollars a year on a quarter acre," she says convincingly, "without using any heating fuel or cooling fans, no grid electricity and no pesticides or harmful chemicals. The greenhouse can produce eighty to one hundred pounds a week, which translates into three hundred people getting five salad servings per week. This could be adapted to a rooftop in New York City, the mountains of New Mexico, or Somalia or Alaska. It could be on the front of a school or community building, on somebody's individual home, or for large-scale commercial production. Harvesting, harvesting, harvesting—it's a very joyous thing to be doing, because harvesting means coming in out of the bitter cold, and taking off your layers and saying, 'Ah, another gorgeous day in the garden.'"

CHICKEN BREATH MOJO

And so were born Anna Edey's "solar-dynamic, bio-benign greenhouses," famed for their mojo magic of chicken breath. She speculates that it may double the plants' productivity. After several years, she gave up on the rabbits, which turned out to be prone to disease and required frequent housekeeping. But chickens require only ten minutes of work a day, which includes collecting the eggs. She found the plants reaching under the wall into the chicken room for the rich fertilizer. "Plants are so incredibly intelligent," she states. "They know just where to go for nourishment."

According to Edey's calculations, each animal puts out about eight BTUs of energy per pound per hour, which gives the heat equivalent to 2.5 gallons of fuel oil per animal for six months' heating. Edey takes a pitchfork to the animals bedding every week or two to aerate the soil, and adds leaves and sawdust occasionally. She empties about 300 cubic feet of bedding once a year and uses the excellent compost.

FAITH, TRUST, AND INSECTS

The most common problem with greenhouses is insect infestations, a condition that causes most nurseries to use large amounts of pesticides and other noxious chemicals. Edey has devised instead what she calls "golden keys" to insect management. "I go through the greenhouse and turn over an occasional leaf. What do I see under there? I see white fly pupae and white flies. I turn over another couple of leaves. Do I see any black spots? The black spot means that *Encarsia formosa* have parasitized the white fly's pupae and out will come an *Encarsia formosa* wasp, a tiny, tiny wasp that will lay its egg in another white fly pupa, and pretty soon you'll have a very small population of the white flies. I don't want to completely get rid of the white flies because then the *Encarsia formosa* would also disappear. I want to have a nice balance between harmful and benign insects.

"If I don't see any black spots, then I place an order from my insectary, Rincon Vitova in California. I order *Encarsia formosa*, and two days later, they are here. I hang up the little cards with the *Encarsia formosa* embedded in them, and they start hatching and go out immediately and lay their eggs."

Edey points out that the common mentality that "the only good insect is a dead insect" is not in tune with biological reality. The huge majority of insects are beneficial, responsible for many of the key biological cycles including pollination. Moreover, insects live in a balance with the rest of life, and eliminating one insect creates an imbalance elsewhere. She sees her job as helping keep the balance.

"I used to use insecticidal soap. I would collect lots of aphids and make bug juice, or make concoctions of garlic and peppers and cayenne. The bottles would get clogged up. Then I started noticing the beneficial insects. I started spraying just individual aphids until I finally said, 'To heck with this! I'm just going to trust that if I provide a good habitat and optimize the environment for the beneficial insects, they're going to do the work. They're not going to do their work if I don't trust them. Faith is a large part of it—letting go, enabling somebody else to do the job."

Edey points out that flowers are very important to the greenhouse ecology. "Flowers provide the nectar and the habitats for the beneficial insects. Some flowers are the favorite of the harmful insects, especially nasturtiums." Originally told that nasturtiums repel insects, Edey found the opposite. But she realized that sacrificing these flowers to the harmful insects allows her to segregate them harmlessly from the other plants.

Today Edey is releasing ladybugs from the greenhouse catwalk. She ordered 35,000 of the pretty red-and-black domed insects, and keeps them in the fridge, where they remain peacefully dormant as if it were winter until she frees another batch. The ladybugs eat harmful insects. She uses a half gallon every six months.

Another Edey golden key against harmful insects is what she calls "living soil." Except for using sterile soil for seed sprouting, she adds a lot of compost, buckwheat green manure, rock phosphate, green sand, and wood ashes to her soil. "Living soil supports millions of critters like earthworms and billions or trillions of microscopic critters. One teaspoon of soil contains millions of fungi and other life forms. We'll never figure out the complexity and marvels of just such a simple thing as a teaspoon of soil."

141

LARGER THAN LIFE

Indeed, Edey recommends that everyone get a small pocket lens to see life more intimately. "If everyone went about with a ten-X lens, and frequently looked at nature, that would go a long way to solving the problems we have because people would come to revere life. Seeing what goes on among the insects is extraordinary—their faces, their legs, their different variations. It's like the bar scene out of *Star Wars*. But they're a lot more intelligent than that. They're enormously perceptive, with sensitivities infinitely greater than ours. A male moth can smell a female moth a mile away and go find her. A bee can smell an alfalfa field six miles away. Through living with nature, we can recall our own latent sensitivity, I believe."

So Edey recommends buying a small five-dollar lens for a Christmas present. "Wrap it up in a big box. There's always something fascinating to look at. A flower opens up your soul. Just as the image gets bigger, your soul and heart get bigger."

Edey is engaging in one of her favorite activities. As she projects the slides onto the wall, she is like a child drawing mischievous mustaches on magazine ads. She has drawn solar-dynamic, bio-benign greenhouse visions across photos of actual cityscapes—high-rise apartments, corporate head-quarters, urban schools. Instead of the glass-and-steel, concrete-and-brick architectural monuments to energy inefficiency and resource dependency, Edey shows her picture of a transformed world. Across a Boston brick building, she shows her retrofit of low-income housing cum greenhouse. "Such an addition could reduce heating costs by seventy-five percent and purify the air," she says, "and give lots of food, lots of bliss." She has drawn a greenhouse across the twenty-third floor of a Manhattan apartment with "kitchen, kitchen garden, hot tub, clean air, and fresh food." Another image is of a large New York corporate tower with a solar greenhouse fa-cade capable of producing ten thousand salad servings a day, 3.5 million a year, plus enough electricity to power 200 homes. Ever creatively uncon-fined, she shows a health spa where the svelte users of stationary bicycles generate electricity while they work out.

"If you borrow money to do this," Edey says emphatically, "you're pay-ing less on the mortgage than you would on the heat and the food you'd be having to buy otherwise. We could have greenhouses, green rooms in which to grow food right in the cities. We can grow most of the vegetables we need in Massachusetts, create jobs, and protect the water supply and soil in California."

Like many pioneers, Edey has received praise, including Amway's environmental award, but found the work slow to progress. "People love it, and people want it. Yet there is little action. So what is it going to take to create the essential change that we have to make? It will require enough truly effective, shining examples of solar-dynamic, bio-benign systems — schools, community centers, conference centers, spas, restaurants, low-income homes — to create a critical mass of informed citizens willing to do what needs to be done to overcome the crippling business-as-usual regulations, codes, laws, and mindsets. It will take enough people to stand up and say, 'It is the right thing to do, and we want it now.'"

Edey is undaunted. She is planning the next phase of her greenhouse revolution. She has honed its economics to a fine art, and has a franchise plan to supply greenhouse designs and marketing plans for entrepreneurs to gross over $200,000 a year on a third of an acre providing their communities with fresh organic produce.

"When you learn how nature takes care of wastes or insect pests, you can simply copy that. But we can also enhance it. I believe we can be very beneficial for the Earth. We now have the technology and knowledge to provide for our basic needs for waste management, food, heating, cooling, and electricity in ways that do not cause destruction of our environment and resources, and that prove that we can do it in ways that improve the quality of our lives and lower the cost of our living. The big question is, do enough people have the will to choose to do it?"

Anna Edey's haunting question fades into the lush scents of the verdant greenhouse. Perhaps the answer can be heard in the breath of chickens.

CHAPTER SIX

"Eco-nomics"

THE NOTORIOUS 1989 EXXON *VALDEZ* OIL SPILL SUFFOCATED PRISTINE Alaskan shorelines under an oily film, saturated seabirds in black goo, choked fish, and contributed positive financial value to the central ledger of our economic measurement, the Gross National Product (GNP). This most environmentally damaging accident in U.S. history added value to the GNP because the $2 billion spent on mopping up the black slick staining the precious wilderness and fishery was counted as positive "income." Similarly, the $40 billion spent annually on treating human respiratory ailments caused by air pollution is welcomed as an asset to the GNP.[1] According to this upside-down system of economic measurement, destroying the Earth is quite productive. Clearly, however, it cannot go on indefinitely.

Behind the curtain of most environmental harm lies an economic motive embedded in a blind-side economic mentality that is flawed in its very assumptions. And because poverty and environmental degradation are inextricably linked (i.e., people cut down forests to feed their families), the widening chasm between rich and poor is itself a potent force of destruction.

If we are to right our Alice-in-Wonderland world of retrograde economics to a biologically based "eco-nomics," a fundamental recalculation of our very assumptions is in order. Such work is already in advanced stages among farseeing economists, companies, and public policy makers. These exciting breakthroughs could realign ecology and economics to bring them into phase with the biological truth of the natural world upon which we all depend.

145

THE PROBLEM

Most of the real wealth in the world either comes directly from the Earth or relies on the Earth's biological processes in various ways for its existence. One conceptual flaw of current economic thinking is that it fails to account for the steady depletion of the finite wealth of the natural world. This myopia is compounded by the neglect of the actual societal costs of products.

Ecological economists have calculated the real value of a single mature forest tree in India at $50,000.[2] Yet when trees are felled, the proceeds are counted as income. This ledger ignores not only the depletion of the resource itself, but also the ecological, economic, and social debits, such as loss of local firewood, erosion of topsoil, silting of rivers, loss of flood prevention and habitat for plant and animal species, and resulting contribution to global warming.

Companies are actually rewarded for environmentally destructive activity and penalized for sustainable efforts by an inverted system of government subsidies and tax incentives. Oil depletion allowances and massive subsidies to the nuclear industry make renewable energy less competitive. Attractive tax and subsidy incentives for logging, mining, and ranching make deforestation profitable. The U.S. government also siphons precious irrigation water to giant corporate agribusiness customers in the Western desert at a financial loss borne by taxpayers and our precious aquifers.[3]

Many products cost society far more than consumers pay directly. The large companies that principally benefit are the recipients of a "corporate welfare" system where the "hidden" costs are passed off to society at large or onto shattered ecosystems. The current system of subsidies and incentives propels corporations in a vicious cycle of environmental destruction.

These transnational corporations are responsible mainly to their shareholders in a relentless system that values the maximization of short-term profits above all else. They are also caught in a savage competition with their rivals. The system compels them to roam the globe in search of cheap labor, weak environmental regulations, docile governments, subsidies, tax incentives and lucrative markets. They are lashed to an internal logic that binds them to this destructive course even when they may want to act differently.

The giantism of the largest of these companies also distorts the rules of the game. According to business critic and author Paul Hawken, "The sheer size of the largest corporations tends to grant them political and economic power to externalize costs that should properly be absorbed by the company, and therefore be factored into the price it sets for its product." In 1991, the ten largest businesses in the world had collective revenues of $801 billion, greater than the world's smallest 100 nations combined. The 500 largest companies control 25 percent of the world's gross output.[4] When people speak about the "free market," one can only wonder if "free" is a verb, or whether we live in a vast "kleptocracy."

Meanwhile, the rich are getting richer by 22 percent and the poor poorer by 21 percent, while the middle class is hardly holding ground.[5] Rich nations now have almost 50 times more real income than poor ones.[6] The same gap operates within the industrialized economies. In the United States, the top fifth of people earn about half the country's income, while the bottom fifth earn a scant 3.6 percent.[7]

One reason for the decline in the income of the huge majority of U.S. citizens is the borrowing spree of U.S. corporations during the 1980s, which by decade's end reached $1.3 trillion of new debt. Colossal corporate interest payments derailed capital for new plants and business expansion. Corporations have limited their share of the tax burden partially through a federal loophole that grants a virtually unlimited deduction for interest on corporate debt. Where corporations paid about 40 percent of total federal income taxes in the 1950s, today their share has dropped to only 17 percent, while taxes have risen by eightfold. Guess who pays the difference?

If these trends continue, the social structure of the United States will begin to resemble a Third World country with a very wealthy elite, a much smaller middle class, and masses of desperately poor urban dwellers. Economic restructuring is essential for both the restoration of the Earth and the social justice needed to create a peaceful society.

Increasingly we also live with a "casino economy" in which money is made from money, not from productivity. Such activity does not produce genuine value, merely a spiraling bubble. Like Las Vegas, the casino economy operates in an artificial world without daylight or real time. This disembodied system is now waking up to the biorealities of the morning after.

SOLUTIONS

Even large companies today are seriously evaluating how to base their endeavors within the limits of natural systems. They recognize that current economic models are simply not viable. Consequently they are starting to address a relationship of corporate reciprocity between society and the Earth.

According to the Worldwatch Institute, a sustainable economy requires "a population that is stable in balance with its natural support systems, an energy system that does not raise the level of the greenhouse gases and disrupt the Earth's climate, and a level of material demand that neither exceeds the sustainable yield of forests, grasslands, or fisheries nor systematically destroys the other species with which we share the planet."[8]

When "full-cost accounting" which accurately values the natural environment becomes integrated into economic thinking, markets will be able to function with genuine efficiency to make ecologically sane long-term decisions. The United Nations Development Program (UNDP) has created the "Human Development Index" (HDI) from three components: life expectancy, literacy, and purchasing power. Ecological economists Herman Daly and John Cobb have devised the "Index of Sustainable Economic Welfare" (ISEW) that accounts for ecological loss as well as the real costs of human welfare.

Economists at the New York Botanical Garden have calculated that the long-term revenues from the sustainable harvesting of fruits and rubber from a small plot in the Peruvian rain forest are twice those from a tree plantation or cattle ranch.[9] Measures like these will permit sustainable planning. Adopting full-cost pricing and environmental accounting methods will limit environmental destruction and improve efficiency.

Another solution is taxing these destructive practices proportionally to their true social costs, while also eliminating poorly conceived subsidies. Such "green taxes" include fees on carbon emissions from burning coal, oil, and natural gas with the goal of offsetting global warming and air pollution. Both Europe and Japan are seriously considering a carbon tax on petrochemicals, which would immediately brighten the economic future of solar and renewable energies.

Green taxes place disincentives on the use of virgin natural materials to encourage recycling and reuse. They impose penalties on the overuse of groundwater. They make producers of toxic waste pay dearly in order to stimulate waste reduction and a safer industrial product cycle.

Ironically, the insurance industry, which has been walloped with more than eight weather catastrophes costing well over a billion dollars each since 1989, has become a strong environmental advocate on behalf of preventing global warming.[10] Combinations of storm events of this magnitude have rarely been recorded, and there are substantial data to suggest a direct link with our industrial practices (i.e., the "greenhouse effect"). Clearly prevention is a far more prudent fiscal option. Such a shift by this hugely wealthy, influential and powerful global industry could change the way the world does business almost overnight.

According to Paul Hawken, the larger solution is a radical conceptual shift to a "restorative" economy. "A restorative economy tries to achieve a market in which every transaction provides constructive feedback into the commons, as opposed to what we know today, when virtually every act of consumption causes degradation and harm. And businesses must—*must*—be able to make money sustaining living systems, or global restoration will never happen. Pioneer restorative companies have survived to this point because of dedicated and spirited customers who have stubbornly resisted economic tugs and pulls and are willing to pay more for their products."[11]

Hawken himself today works closely with the Natural Step, an environmental education and training program for corporations which teaches the principles of eco-nomics and its efficiencies and values. Companies big and small are employing the program and beginning to implement elements of it into their operations. As large companies get involved, their very participation alters the currents of commerce and influences competitors to respond.

There are growing numbers of companies, both large and small, acknowledging the folly of current business practices, and redefining industrial ecology and "intelligent products" (see Chapter 7). Many consumers are voting with their pocketbooks for these companies with genuine green strategies and products that reflect a social or environmental "premium." Other innovative approaches such as "debt-for-nature swaps" forgive Third World debt in exchange for preserving ecosystems.

It is not to say that these changes will be quick or easy. It has taken a long time to reach today's crisis, and it will take a generation or two to reverse and undo it. The Navajo Indians are designing a 500-year plan for their people and land. The founder of Sony constructed a 500-year business plan and carried out the first 60 years himself. These are the sorts of time frames to which we need to adjust.

"Eco-nomics" also forces us to reevaluate our patterns of consumption (see Chapter 8). It is time to discard our societal identity as "consumers" to reclaim our role as citizens of the ecological commons in a reciprocal balance with the natural world.

Unfortunately, however, very little is being done to narrow the gap between rich and poor. This painful situation will almost certainly undermine these other positive trends, if it continues unattended.

Nor do these solutions address fundamental issues of corporate tyranny, democracy and sovereignty. In his pamphlet "Taking Care of Business," author and activist Richard Grossman has described a practical means of challenging corporate domination by revoking corporate charters, the legal instruments issued by states which grant them their legal standing. This legal tool is a legitimate and interesting means of restructuring corporate power, but it is unlikely to have impact without a significant political movement behind it.[12]

WHAT YOU CAN DO

As individuals, we need to take personal responsibility for our livelihood and how we use our money. Question the impacts of your consumption and act accordingly. Many people are engaging in "right livelihood" by seeking or creating positive jobs for themselves that follow their passion and serve a real need. Numerous socially responsible funds now exist which use social and environmental screens for their investment capital (see Resources Section). Shareholder activism is having real impacts on how companies do business. Above all, we need to stay connected to our money and avoid blindly turning it over to the vast corporate system of amoral destruction, which values only short-term profits.

The innovative solutions described in this section point the way toward this new "eco-nomics" based on respecting natural limits, demanding corporate reciprocity, and investing from the heart. Joshua Mailman and Jason Clay are lighting the way through the dark recesses of money and business toward an equitable future that honors trade but also respects people, life, and spirit.

Joshua Mailman's
Green Money Laundering

Joshua Mailman has spearheaded the financial community in supporting capital investment for socially responsible businesses. He has helped move large sums of money into social ventures and has jointly created organizations which actively seek to redirect capital and engender profitable businesses which serve the common good, their employees, and the Earth.

THE HOTEL BANQUET ROOM IS THE USUAL DEAD-AIR, OVERDECORATED, plastic extravaganza with one set of business conventioneers after another passing through its modular soul. But gathered tonight in the Atlanta hotel is the Social Venture Network (SVN), a most unorthodox group of businesspeople, social entrepreneurs committed to doing well by doing good. The room includes public figures such as Anita and Gordon Roddick of the Body Shop, Ben Cohen of Ben and Jerry's, and Wayne Silby of the Calvert Social Investment Fund, along with hundreds of the most progressive, mission-driven companies and nonprofit activists in the country.

Anita Roddick, outspoken founder of the $600 million Body Shop, a personal-care products company whose commercial success has been built on its strong stand on social issues, introduces the keynote speaker affectionately as "wonderful Josh." Josh Mailman has stayed up most of the previous night penning his speech. He peers over his round-rim spectacles with bemusement, scratches his head, and snorts like an animal pursuing a scent. As cofounder of SVN, he has seen the organization grow from an initial gathering of 70 friends only six years ago to over 500 members today. He has personally invested in a high proportion of the companies present, and donated generously to many of the nonprofits. He has played a seminal part in launching a revolution in the redirection of global investment capital toward social goals, and it is obvious that it comes from his heart.

"MONEY IS LIKE MANURE"

Mailman braces himself on the podium to lift off into a typically expansive vision of a new society based on economic justice. "My father always told me," he begins wryly, "that money is like manure. If you pile it up, it stinks. But if you spread it around, it can do a lot of good. If we do not use our wealth and resources to help lessen the gap between rich and poor, when the world looks back on us from the twenty-first century, people with money who didn't do anything will be remembered as the war criminals of the nineties." It is hardly the sort of rhetoric endemic to conventional capitalist soirées like the notorious Predator's Ball. The room thunders with applause.

Over the course of the weekend, Mailman is a grasshopper alighting briefly in room after room, doing deals, striking alliances, introducing people, casting an impassioned plea for a favorite cause, conveying his vision, and listening closely to the vision of others. In this fertile culture of activist commerce, he is a human pollinator with a prescient knack for connectivity.

They say that if you want to know what God thought of money, look at the people she gave it to. The selfish pursuit of material wealth has swollen to become the dominant value of society. Business, the main motor of human activity, today counts the bottom line to the virtual exclusion of other values, while most people expend their lives in a struggle for economic survival. "Those that have shall get, and those that don't shall lose," as Billie Holiday wailed, but the blues singer could scarcely have imagined the "greedlock" that consumes late twentieth-century life. The concentration of wealth today is the most extreme it has been in modern history, as an estimated 3 percent of the population control over 80 percent of the world's wealth.

Mailman is a notable exception to this gnarly picture, exhibiting a long record of putting his money where his values are. His passions span social and economic justice, environmental restoration, women's rights, natural medicine and a host of other compassionate and visionary endeavors. This philanthropist, private investor, and adventure capitalist runs a sort of green money-laundering operation, helping fortunes and those who possess them to come clean.

In his modest offices on New York City's upper East Side, the phones at Mailman's Sirius Business Corporation ring like an endless chorus of electronic crickets signaling a never-ending chorus of prospective transactions. He is an emerald magnet for the impassioned dreams of social venturers, as well as all the crazy hustlers. Mailman breaks one conversation off in midsentence to take another call, arcing across distant synapses to bridge another far-flung deal. Like Alice's White Rabbit, he is in perpetual motion scampering to the next date.

DOING WELL BY DOING GOOD

"We've really lost an idea of the commons," Mailman says, shaking his fuzzy head earnestly, "the idea that there was a common ownership, a common stewardship, a common heritage. For thousands of years—or for traditional cultures, tens of thousands of years—we were not dealing in a global society that was connected and trading all over the world. That advance in connection has not led to a greater stewardship, but rather to more taking from the commons. The accumulation of capital in individual hands or in one country's hands has become the name of the game, driving ecology and cultures into the ground through the exploitation of resources and individuals.

"There is almost a united front of many companies," he continues, "that are not willing to rethink how they're doing things in a substantive way quickly enough in order to avert significant increases in the loss of biodiversity, loss of cultures, loss of wilderness, loss of the very underpinnings of civilization. We've had the illusion that we're making 'progress,' but if you took basic indices of human suffering, we're really deluding ourselves. This idea of commercialism as a goal is killing us. Instead of investing in successes, we're investing in disasters like nuclear energy, giant hydroelectric plants, and poison-based agriculture. Just from an economic point of view, this is not a good way to run the world. If you don't invest in solving problems early on, then you invest in the disasters that result. You don't even have to be concerned about human rights to see that kind of logic."

Indeed, Mailman's scenario is rapidly being borne out today. Economic analysts are finding that companies that do the right thing have a comparable bottom line to standard investments.[13] Some investment funds that use social screens such as environmental responsibility, minority hiring, and community involvement are producing higher profits than conventional funds. A recent study by the Institute for Southern Studies, a social policy research and advocacy group, concluded that "nearly all of the [U.S.] states that ranked among the top dozen using the study's environmental criteria also ranked highest using its economic criteria." According to an account in the *New York Times*, "'At a policy level, the choice is really not jobs versus the environment,' said Bob Hall, research director of the institute. 'The states that do the most to protect their natural resources also wind up with the strongest economies and best jobs.'"[14] Mailman sees this process taking place at the global level as well.

Mailman started consciously scanning the private sector for companies whose very businesses are grounded in ecological principles and social justice. "I thought that there were going to be lots of possibilities in new start-up businesses that were not only good business ideas, but that would be run differently and would themselves be able to contribute to creating a more just and sustainable society. I had real interest as a private investor in putting my money where my mouth was, and saying, 'How do we invest in new ways of doing things?'"

Mailman's capital is wired into many of the strategic circuits of the bioneer revolution. Although his myriad leading-edge investments have doubtless produced years of intensive psychotherapy for his financial advisers, he has regularly taken an early lead investing in companies he believes are serving the social good. Raising start-up capital for visionary companies is just slightly easier than raising the dead, and his early-stage financial support is often decisive.

One business Mailman helped support is Energía Global, an enterprise bringing solar and other energy-based appropriate technologies to Latin America. The company produces both jobs and inexpensive, environmentally sensitive power for communities that would otherwise be without it. Another investment was Shaman Pharmaceuticals, which is protecting biodiversity by developing pharmaceutical drugs from higher plants from the rain forests based on the knowledge of indigenous peoples (see Chapter 4). Mailman has also funded a green technology promoted by John Hay, an early principal in Celestial Seasonings teas, whose new company sells a "green plug" to cut energy usage by 10 to 15 percent on old refrigerators and freezers. In four years, the venture is doing $10 million of business, and now has utility companies involved in its ownership. "When a utility company can make money investing in a company that's saving energy," Mailman grins, "then they're going to say, 'Hey, I'll do more investing like this.'"

All these companies come riddled with imperfections, limitations, contradictions, and paradoxes. They must operate and survive by business-as-usual rules, which are set up against serving values besides the bottom line. Because they hold themselves to a higher standard, they are often viewed more critically, as the Body Shop experienced in 1994 in a widely publicized derogatory analysis. All regard themselves as working models, and make little pretense about falling on their faces from time to time. But at least they are trying. The sobering fact is that nine out of ten new businesses fail, and unconventional business ventures are even riskier.

"I haven't only looked for good ideas," Mailman says philosophically. "I'm looking for the people who can execute the ideas. I'm looking for people who have a really strong vision, especially in areas that other people might not have been involved in. With some of these, there was no assurance that they would succeed. There were certainly bunches that failed."

THE SOCIAL VENTURE NETWORK

Mailman's personal calling did not stop with his own bank account. Acting as the grain of sand in the oyster of the financial community, he has prodded, provoked, and inspired many others to join together into a greater social force to transform the economic system. He teamed up in 1987 with long-term ally Wayne Silby, founder of the Calvert Social Investment Fund, a pioneer in social investing that today manages a portfolio around $1 billion. Together they brought together a group of friends from companies, nonprofits, and the investment and activist communities. The outcome of that meeting was the formation of the Social Venture Network.

"It became apparent after the first SVN meeting that there was the possibility of creating a new network of socially committed businesses. There was a desire among the founders of many companies for community beyond their individual trade organizations. A lot of people, many of a generation that grew up in the sixties or seventies, had been influenced by progressive ideas. Early on they had come to a point where they said, 'I want to be a part of changing the world.' They had an overview of how many things were going wrong and saw a connection among those things, whether it was environmental destruction, social justice, an appreciation for native cultures, or an expanded awareness of the inter-connectedness of life. They said to themselves, 'I'll never let go of this. This will always be with me.'

"SVN has really been a kind of call," Mailman continues, "a gathering point to say, 'Okay, don't forget what you said to yourself twenty years ago. Now is the time that we've got to inspire and challenge one another.' It's not about 'The one who dies with the most toys wins.' It's about what is left for future generations and how we can strategically work together to bring that about."

The Social Venture Network is indeed a network, not an organization with a coherent or fixed point of view. Many of the companies are characterized by a visionary founder who had a mission, and often making money was a secondary motivation to fulfilling a social goal. The Earthrise Trading Company, now the world's largest producer of spirulina, an algae used for health food, which was recently documented for its anti-AIDS activity, was started by Robert Henrikson to bring a sustainable health-food product to market which can be grown even where there is no farmland. John Schaeffer founded Real Goods Trading Company to make solar and appropriate technologies commercially available in the United States. Progressive Asset Management started as an alternative for investors who did not want to invest in stocks of companies engaged in negative environmental and social activities. PAM has promoted the "double bottom line" of both financial profitability and social justice.

The issues raised by SVN companies often revolve around how a company treats its employees and how the company's vision can be better reflected in its internal relationships. Another focus of socially responsible business is the product itself, what it is made from and how it is produced. Yet another concern is how the company relates to its community, whether it only takes or also gives back. Many view this concept of corporate reciprocity as the single greatest failure of conventional business. SVN has produced a survey of the "Best Practices" of member companies dealing

157

with a wide range of such issues, including employee stock option programs, relationships with compatible nonprofits, vision statements, and hands-on management tips.

The Body Shop was an early player in SVN. The business had grown tremendously based on its use of natural products and its pioneering social-issue marketing. The company has regularly conducted campaigns against animal testing, has supported AIDS education, and has tried to procure products from the rain forests sustainably harvested by indigenous peoples. Another key player in building SVN has been Ben Cohen of Ben and Jerry's Ice Cream, which has paid premium prices to support local dairy farmers in New England, as well as Native American blueberry farmers.

"Despite the high fat content of the ice cream," Mailman muses, "I have continued to be extremely impressed with the values and integrity that Ben Cohen and Jerry Greenfeld have brought to Ben & Jerry's. They have consistently tried to move the goal posts. They are now doing a program about diversity issues. There are not a lot of non-white people living in Vermont, but it's their commitment to explore issues about how men and women, gay and straight people, and people of different backgrounds can maximize their collaboration together. They're basically saying that if you're sexist or racist or homophobic, there's going to be less space for you in companies than in the past. That kind of commitment is very important because those are the same kinds of problems we're dealing with in global society, and can be a really positive influence that America could have on the world, because we have such a culturally and ethnically diverse society."

GROWING A COMMUNITY
OF BUSINESS ACTIVISTS

Mailman sees SVN as an incubator for growing the community of alternative businesses. "We don't have enough leadership in the world with any kind of vision. In SVN we have brought together people who are leaders in their own communities and who are looking to take more leadership. One of my personal concerns is how businesspeople become social activists and how social activists become businesspeople. You have social activists, people with brilliant ideas spending half their time trying to raise their money so they can get their projects off the ground. That's a pitiful way to run a system. They're the ones that for me represent the best of humanity. They are the ones that we should be going to and asking, 'Where are we going? What should we be doing?'"

SVN and its members now appear to be growing into the role of pilot fish to the larger business community. Very large corporations and government institutions are examining SVN successes, and exploratory discussions are underway with several of these on how better to address both social and environmental goals through business. What, Mailman wonders rhetorically, will happen as large companies find that they can improve their bottom lines by improving their environmental performance and granting employee ownership? What would happen if the U.S. government, the single largest buyer of goods and services in the nation, applied sustainability criteria for its purchasing? These are the questions that Mailman ponders and releases like swarms of butterflies to pollinate the SVN community.

It is notable that SVN companies receive far more publicity from their social-action programs than their meager advertising budgets could ever buy. Mailman believes that it is a function of their doing good things in the community. "It's what is called 'stakeholder relations,' which in this case means cause-related marketing. It's not a marketing campaign. It's saying what the things are that we have in common with the community. There will always be Genghis Khans running companies that can be successful as well, but on a long-term basis I think that people want to feel that there is a vision where they work, where they're spending a major chunk of their waking hours. Companies that have a real vision, a purpose that has some sense of a spiritual value to it, will be stronger companies over time because they will reflect an idea that society is not here to serve business. Business is here to serve society."

Mailman foresees a blossoming of companies run by women and minorities. "Whenever you give people a chance who haven't had a chance, you can really up your odds of success because people want to lead decent lives. I really believe that people are ready to work for that, but they can't work if they don't have any tools. If capital is put into the hands of the people living in the villages of the world, people that haven't had a chance, you'll see real productivity. We have an incredibly wise and creative human community out there waiting to be drawn on, waiting to be included, that has been ignored by the international institutions in a very significant way.

THE MOST IMPORTANT ASSET: THE VISIONS OF THE PEOPLE

"Everybody has dreams," Mailman continues softly, "but people bury their dreams when they don't think they have a chance. We have to care about the dreams of the people in our societies, in our cities, in little tribal communities around the world. We have to ask them how they want to live. Everybody has a right to be a part of the decision of how they live. Otherwise we promote a slave society. If you have economic slavery, you have a system that is wasting its most important asset, which is the visions of its people. We need all of the vision and all of the dreams that we can get in the world because we've figured out how to sell a lot of stuff, but we've been losing the ability to dream. The dream has got to be something besides a commercial dream. The dream has to be a dream of the great teachings of the world. What are the treasures of knowledge that have been accumulated over the millennia? What are the dances and the arts? What are the plants and the languages? What are the things that really collectively comprise the jewel of humanity in all its diversity?"

To realize these dreams access to capital is crucial, and Mailman points to the model of the Grameen Bank in Bangladesh, which is now successfully lending $500 million a year in small amounts to the poorest people of Bangladesh to form cooperative economic enterprises. A similar program is the Women's Bank in India, which discovered that making small loans up to 100 dollars directly to women yields phenomenal results. Women, who had previously been financially disenfranchised, have in fact proven to be the social glue of the community. Given access to capital, they have succeeded spectacularly. They hold community meetings to assess cottage industry loans, and dispense the small loans for family-based projects, such as weaving and farming. The payback rate has been an astonishing 98 percent. Similar programs are now being successfully replicated worldwide.

Mailman is enthused about a newly completed joint venture with the Grameen Bank to bring cell phones through his wireless communications company to the poorest of the poor in Bangladesh. Grameen (it means "village") Bank is "owned" by its 300,000 members (and 10 percent by the government), and has become world famous for providing "microcredit" to the poor who are otherwise shunned. In addition, it lends to 93 percent women, never more than a hundred at a time, and its repayment rate is also 98 percent, a figure never thought possible by bankers. Mailman's wireless joint venture will now bring cell phones to a country where only three people out of a thousand even had access to a phone, facilitating many business opportunities for rural development among the poor.

A WEALTH OF OFFSPRING

With Mailman's stimulation, SVN has spun off two important offspring. The Investors Circle seeks to place capital in socially responsible companies. In its first three years, members of Investors Circle put an estimated $10 million dollars into such ventures. The other SVN progeny is Businesses for Social Responsibility, a national chamber of commerce for companies wanting to express and adopt the kinds of values exemplified by SVN principles. BSR now has over a thousand members in ten regional chapters.

Mailman was born into a wealthy New York family whose self-made fortune started with his father selling straight razors door to door, and grew into one of the first conglomerates. He could easily be sitting by the pool with his cellular trading options, but he was moved by the social revolution of the 1960s. "I saw that commercial interests were more important than any kind of moral standard behind the government. I knew that I would try to use my own time and resources to bring about change in that area and try to make the world a better place."

Mailman galvanized the formation of a new foundation in 1981 called Threshold which sought not only to fund spiritual, humanistic, environmental, Native American and other largely neglected areas, but also to help its members come to terms with their wealth. While most people fantasize that inheriting money would be great good fortune, it usually uncorks a plethora of problems and identity crises as well. Threshold became a safe haven to process the transformation, and ultimately to help people become better at giving. Possessed of an urgent drive to fund social reforms, Mailman quickly became notorious for his relentless networking and almost prosthetic phone presence.

Threshold has become a model of innovative grant making in the foundation community. Today the foundation gives out over $1.5 million a year, often to the most neglected, controversial projects, which would otherwise be orphaned.

GIVING BACK A COMMON HERITAGE

Subsequently Mailman recognized global business as a primary means for social change, and turned his efforts to revolutionizing how business is done. He is characteristically modest about his role in advancing the ethical financial movement. "I'm much more sanguine about the impact that we've been able to have, but I also don't want to discount small acts. We have a need for small acts, and I consider the things I've done small acts, hopefully, compassionate acts. To the extent that I've been able to make a contribution, it's been out of a desire to build community, realizing that I'm no more important—and I think in many ways less so—than some local activist. The real leaders are the people that are in there day after day, slugging it out, who have chosen something other than monetary gain, who are there because they are fed by the experience of community that they have."

Mailman believes that people can come to terms with money as a force for positive social change, whether they have a little or a lot. "One thing that I would say, particularly to people that find themselves in a situation of having wealth in a world of need, is don't be afraid to act. Ask whether your allegiances are simply toward the preservation of capital in this old-world model of passing on more and more money to the next generation. When your children or grandchildren say to you, 'What did you do to change the world? What did you do to really make a difference?,' you can say, 'I took our individual heritage and made it into a common heritage. I took what had been taken and gave it back.'"

The coming years will bring a tremendous transfer of wealth of about a trillion dollars to Mailman's generation. "It's very conceivable that we're in a time where we'll look back fifty years from now and say, 'If only we had done that, if only we had made it a priority, then we could have saved this.' It's not a time to stand back and be disempowered. It's a time to take action. Let's redress the imbalances that have come out of a civilization that has not had human needs at its core. Those small acts of courage are the only thing that will get us through."

The phones at Sirius Business have finally gone dormant as the evening descends. Mailman is softly reflective as he gathers his coat to venture forth to a benefit concert for the rain forests. "There's a line in the Talmud that says, 'It is not up to us to complete the task, only to do our part.' I think that we really have to look at our lives and say, 'What's our part? What is it that we were meant to do?' That will draw people who have found themselves in positions of having significant resources into a common humanity. It's probably easy to live with lots of houses and lots of cars and lots of stuff, but is that really what's going to make you happy? Is that really what's going to put you in greater touch with your soul?"

Mailman ruffles through the papers distractedly. He is contemplating participating in a venture fund to capitalize a Biodiversity Fund. There's a large venture pool to revitalize Eastern Europe, emphasizing environmental technologies. He is funding "Ain't Nuthin' But a She Thing," a two-hour TV music special on the lives of remarkable women. He's heading south to wire SVN Latin America. Then there's the plan to create community-based seed banks of traditional seed stocks in India.

There is certainly no lack of work to do, and as Mailman dons his coat against the winter cold, he taps his foot in anticipation of the hot beat of the rain forest throbbing inside his heart. "I think of the quote from the great Indian poet Rabindranath Tagore, who wrote, 'I slept and dreamt that life was joy. I awoke and saw that life was service. I acted and beheld service was joy.' How we can get to that point of selfless service cuts to the heart of what it means to be a human being. The people who have most inspired me along the way have had that quality. That's a great goal for people who have power because it's the transformation of the love of power into the power of love. That's what I look to continue to grow in myself."

Jason Clay's Social Trade

Jason Clay has devised innovative economic strategies for biodiversity and cultural conservation. His methods give added value to sustainably produced products, and help support traditional cultures and local people.

PACING THE RIVERBANK ON THE BRAZILIAN AMAZON, JASON CLAY'S EMOTIONS are turning as red as his hair. There is a 1,000-ton load of Brazil nuts on a docking barge. They will be hauled nearly 3,000 miles away to be shelled. In the process, one-third will rot and be thrown away, and not one cent of value will be added in a region with 50 percent unemployment.

Walking on top of the nuts, Clay knew there had to be a better way. That day he placed an order for 15 metric tons of Brazil nuts from a local factory and began to explore how to set up shelling operations in the forest and poor urban areas where people were starving for jobs. "How did I get myself into this?" wondered the anthropologist turned commodities broker.

Thousands of people later woke up across North America and drowsily poured their breakfast cereal. They perused the cereal box of Rainforest Crisp, which told the story of natural ingredients sustainably harvested from the South American rain forests. Part of the profits were going toward helping save not only the endangered jungles and its riot of biological life, but also the peoples who have lived in balance for eons with their now threatened forests. The message was from Cultural Survival, the nonprofit group whose marketing program Clay founded.

"CONSERVATION IS A PEOPLE ISSUE"

"Conservation is a people issue, not a biological one," says Clay, leaning back in his chair at the World Wildlife Fund office in Washington, D.C., amid mounds of papers and economic analyses. "Trees don't cut themselves around the world—people do. We're talking about conservation, not preservation. It's not an option to build fences around the world. Every bit of rain forest is claimed by somebody. Every bit of rain forest has been altered by people, whether they're indigenous people, colonists, or others. It's really a question of using resources wisely, not building fences around them."

Shifting his red-rimmed glasses, Clay continues passionately. "Green business has focused on the symptoms of environmental problems, on the packaging of products and recycling. But what's inside the package? How is it created? Harvested? Cultivated? Managed? Is it sustainable? The causes we need to look at are population pressure, poverty, greed, and ignorance, and a special little subset of greed I would call the overconsumption of the world's resource base."

The TV droning in the background is tuned to CNN's report from the New York Stock Exchange. Clay takes the cue. "What do we value? Every day on the major networks you can find out what happened on the New York Stock Exchange, how many hundreds of millions of shares sold down to an eighth of a cent in value for billions of dollars' worth of shares exchanged every day. You can find out what the price of gold did through hundreds of thousands of transactions in six or seven key locations on the planet. We clearly value this knowledge. But we don't know how many species there are on this planet, and there aren't as many species as there are stocks traded each day. On the human side, we don't know how many homeless people there are, or the cost of feeding a family, or the infant mortality rate. We don't know because we don't care. We need to change the way we think about business and economics."

According to Clay, a major part of the problem lies in consumption itself and the severe strain it is putting on the world's resources and peoples. United States citizens consume per capita 20 times more than the rest of the world combined. The world simply can't support such an extravagant lifestyle for the vast majority of people. It is therefore essential to educate consumers on where products come from, what it takes to produce them, and how it is done, with the goal of changing consumption patterns.

"MAKING A KILLING IS KILLING THE WORLD"

At the core of the problem are the values of our economic system, Clay says. "In the last thirty years business has shifted from making a decent profit to 'making a killing,' and the killing in the market place is killing the world. Income, prestige, and promotion are based on short-term goals. Quarterly reports are never going to be a way to manage the Earth's resources. You can torture the data to create any kind of quarterly report you want. Stock manipulation and junk bonds are not a productive activity, and don't generate real income for real people who are sweating. Corporate takeovers are not productive activities."

We need a longer term view of self-sustaining activities, Clay contends. He tells the story of a meeting in Arizona between Hopi Indians and the U.S. government to discuss a mining company that was protesting the regulatory demand that it cover over dangerous uranium tailings with a foot of asphalt. The government representative said that the company was not going to be around in 10,000 years, the federal time requirement for safe disposal of the radioactive waste. He further speculated that even the U.S. government wasn't going to be around in 10,000 years. A Hopi stood and replied, "But we're going to be around in ten thousand years," a reasonable time frame for indigenous peoples.

THE HIDDEN FORCE BEHIND ENVIRONMENTAL DESTRUCTION

According to Clay, it is the world trading system itself that is destroying the Earth. Clay offers a "history lesson that you never learned in school," and it may well represent the most astute analysis in the world today of the hidden forces driving environmental degradation.

Most of the political states of the world have been created since World War II, he explains, and these new political entities have no legitimate historical basis. Some 150 of the nearly 200 states today part of the United Nations arose just since 1945. Since 1945, these states have written three thousand constitutions, which have little to do with democracy or justice but much to do with allocating the ownership of resources. Yet these political states contain within their borders some 7,000 nations, cultures which have existed for hundreds or thousands of years. Nations are groups with a recognized territorial base, a language, a history, and often a religion. Most importantly, they have a history of governing themselves.

166

Most states are, in fact, multinational empires. They are ruled from the top down. These states were established to accomplish ostensible goals of creating a measure of stability and keeping global wars at a minimum. Instead, they have spurred civil wars, battles between states and the nations within.

Above all, the state system allows for the free flow of trade—minerals, oil, timber, land, water, hydropower, and now intellectual property and genetic materials. The trade allows developed countries—principally the United States, Europe and Japan—access to the world's finite resources, which are mostly located in the Third World countries, without running costly and complicated colonial systems to rule them. The new Third World states concentrate power in the hands of a few elites, dictators, and single-party states who are our trading partners, who supply us with the resources upon which we depend to maintain our standard of living. In most of the world, government is the biggest game in town, Clay adds.

The system works by instituting one local group in power, says the anthropologist who has lived and worked on five continents studying ethnic conflict. In Ethiopia, this strategy meant putting the Amhara people in power till recently even though there are 80 other ethnic groups in the country. In Kenya, the Kikuyu ruled over 100 other nations in the country. Elites adopt a winner-take-all strategy. Developed countries prop up these elites through development assistance, military support, and foreign investment. The elites fix commodity prices, appoint commodity boards, and control commodity export into the international trading system. These revenue sources account for two-thirds of all the revenues of Third World states. The rest comes from taxes.

When the elites borrow for investment, development, or arms, everyone is forced to repay the money. The elites don't suffer from the staggering Third World debt. To the contrary, the elites have foreign assets tucked away in Swiss banks, New York real estate, and elsewhere that are equal to the entire Third World debt.

What is causing Third World debt? Military expenditures are equal to half the debt, for weapons needed to maintain the elites in power. In Africa, military expenditures equal the entire African debt. Weapons are directed against internal enemies. More than 75 percent of the 120 shooting wars in the world today are between states and nations. Known in international parlance as "low-intensity conflicts" because the superpowers are not involved, these internal wars have resulted in over 5 million deaths, mostly of indigenous peoples. At a given moment there are 15 million refugees and 150 million people displaced from their homelands who haven't yet fled across an international border. "These people are not suffering from acts of nature," Clay says with outrage. "They're suffering from acts by people." The same acts come from dictatorships of the left or right, or religious or sectarian states.

But once indigenous peoples are displaced from their homelands, they are jettisoned from the lands they know how to manage, leading directly to environmental degradation. "You can take a hunter-gatherer group," says Clay with anguish, "move them forty miles, and they will starve to death because they don't know the resources of the new location. You can take an agricultural group and move them twenty miles, and they will do untold damage to the environment because they don't know how to farm it. That is what one hunded fifty million people are doing today."

The major cause of the conflicts is natural resources. The states, which make money by selling resources, appropriate them from indigenous peoples, leading to conflict and armed struggle. Weapons' purchases lead to debt, and in turn to the need to appropriate more resources to pay for the debt. The vicious cycle stems from international trade.

"ENVIRONMENTAL DEGRADATION AND HUMAN RIGHTS VIOLATIONS GO HAND IN HAND"

The formerly self-sufficient indigenous peoples are displaced, disenfranchised, impoverished, and disconnected from their homelands. "From the former Soviet Union to South Africa, from Brazil to Burma, from Mexico to Malaysia," Clay says, "they are saying, 'Enough. Stop the theft of our resources. Leave us alone. Without the resource base we don't have a future. Without us, the world doesn't have a future.' Indigenous nations know that their future is tied to their resource base. They want to control the factors that affect their future. The nation agenda is land rights.

"Environmental degradation and humans rights violations go hand in hand," Clay goes on. "You can't cut the trees until you get rid of the people or deny the rights of those people to the land and resource base. In Brazil since the turn of the century, one indigenous culture has disappeared each year, ninety out of two hundred and seventy nations, while twelve percent of the forests there have been destroyed in the last four hundred years. Human rights violations—the loss of cultural diversity—precede the loss of biological diversity. Environmental problems are not going to be solved unless we have social and economic justice."

Consequently, protecting the environment is inextricably linked to protecting the land rights of the people who live there and have used these systems without destroying them for hundreds or thousands of years. "People degrade environments when they don't have options. The goal has to be to help people who live in fragile ecosystems make a living without destroying those resource bases. That can mean finding markets, crops, or ways to get more value from what they're producing—securing resource rights and legal rights as citizens."

Clay suggests that life in the rain forest is hardly romantic. Four out of six children die before the age of five, and average life expectancy is only 40 years. Indigenous people do not want to live just the way their ancestors did, but they do want to maintain many values and traditions. They want to make a choice.

"This is the backdrop for how an anthropologist and a human rights organization got involved in selling commodities from the rain forest," Clay muses at the irony. "We're trying to use a trading system, an existing consumer demand, to fundamentally change a trading system and forever alter consumer demand."

SOCIAL TRADE IS BORN

The idea of social trade originated at a Grateful Dead concert where Clay met Ben Cohen, cofounder of Ben and Jerry's Ice Cream, who wanted to help stop rain forest destruction. They came up with the idea of a rain forest ice cream with ingredients purchased from sustainable sources and forest-based peoples. At a large gathering in the Amazon to stop a massive hydroelectric project, Clay got similar interest from Anita Roddick of the Body Shop, who wanted to apply the strategy to her body-care products company.

169

Clay had worked for ten years for Cultural Survival (CS), an academic Harvard-based anthropology group. He studied ethnic conflict around the world, especially in Africa, and founded *Cultural Survival Quarterly*, where he wrote on the declining state of the world's indigenous peoples. "We needed to figure out how to predict these disasters, find the causes, and prevent them. Our role ought to be to get out of the body-bag business and get into helping people make a decent living. For one book, I wrote about how one of the best things you could do is start a market." Finally he decided to step through the anthropologist's lens, an academic taboo, and form a non-profit marketing arm of CS called Cultural Survival Enterprises (CSE).

Initially, he bought rain forest commodities on the open market to get some cash flowing, since not a single forest-based producer group existed at that time capable of export-quality products. He attached a 5 percent premium price to the products, and negotiated profit-sharing arrangements on the finished products with each company. Five years later, the marketing program was doing $3.5 million of sales. In addition to the commodities traded, around $1 million of revenues was generated from environmental premiums, profit sharing, and foundation grants, which went to the producer groups and regional human rights organizations.

Using publicity to create demand, including feature articles in national magazines such as *Fortune*, *People*, and *Newsweek*, he soon had 200 companies on line to develop products, and was exploring 350 commodities, later narrowed down to about 15. Within three years, 11 of the commodities were being produced totally by the local communities. The project resulted in the first factory owned by the people who collect Brazil nuts, allowing export to the U.S. of shelled, dried, vacuum-packed nuts. Instead of getting 3 percent of the New York price, they now get 65 percent. An estimated 35 million people in the U.S. saw the packets and boxes that had information about the rain forest and its peoples, highlighting the power of consumers.

Clay notes that the 5 percent premium translated into less than a penny on a four-dollar box of cereal. The problem with conventional business accounting practices is a series of exorbitant markups by everyone who touches the product, Clay found.

This approach has been used to provide practical, ground-level solutions for several forest-based groups. The Huichol Indians, who live in the

Sierra Madre mountains of northern Mexico, were living a subsistence life augmented by timber sales to Mexican logging companies. The logging, however, was depleting the fragile desert mountain ecosystem, and the Huichol approached Cultural Survival for advice. They followed the group's suggestion to make a value-added product: furniture for sale on the local market. Now they are getting 300 dollars a log instead of about one dollar, cutting a tiny fraction of the logs they formerly did, and achieving viable reforestation. Other groups learned techniques of grafting fruit trees and producing specialty crops in a much more rapid production time frame.

Clay notes that CSE dealt only with groups which were already in the market or wanting to become engaged, and he acknowledges that the culture shock and money economy can also create problems. "Not every community is going to benefit, and not everyone in every community wants to get involved. Our point wasn't to drag indigenous people kicking and screaming into the market economy. Our point was to say that ninety-nine percent of the seven thousand groups around the world are already in the market to some extent. Most want the benefits of trade. Their situation can be improved by negotiating better relationships."

But Clay quickly became a victim of the social trade program's success. Cultural Survival's original academic founders buckled at this growing involvement in real-world commerce, and after a turbulent three-month debate about the program, abruptly canceled it. Clay found himself out on the street, the program's $4 million bank credit line streaming off without ballast. He landed on his feet at the World Wildlife Fund, the world's largest international environmental organization, and founded another group, Rights and Resources, to process the lessons learned and move the work to the next level.

COMMODITIES AND THE ENVIRONMENT: BLIND-SIDE ECONOMICS

Clay came to see the work of Cultural Survival's marketing program as isolated and very small in the larger global picture of macroeconomics. "These top-spin kinds of commodities like Brazil nuts used in Rainforest Crisp weren't really having much of an impact on global trade." That's when

Clay threw himself into the money pit of commodity trading, and that's why he is on his way next week to Belize to negotiate the purchase of a large sugar plantation.

Production of the world's 15 or so major commodities—coffee, sugar, rubber, soybeans, wheat, bananas, corn, cotton, and so forth—occupies 30 percent of the arable land we use, and they are the most polluting human endeavors. In fact, the most profound effect of human activity on the environment may well be through the extraction or production of raw materials for major commodities used in the economic process, according to Clay's research. Yet remarkably, he found, few have ever examined this situation, much less tried to correct it.

To begin with, natural resources have been treated as free goods, their depletion not factored into the balance sheet. The price of corn and soybeans, Clay illustrates, should be partly based on what it takes to maintain a river system in the Midwest. The environmental costs of such production were relegated to an "externality," a factor not considered as part of the economic equation. The costs of massive flooding in the Midwest caused by unsustainable farming methods, which has cost $20 billion to start to repair, have been borne by the government and taxpayers, not by the producers.

As companies became larger and more powerful, they sought control not only over the natural resources, but also over production and markets. Their sheer size reduced production costs, and they substituted machinery for labor in larger and larger production units that were vertically integrated to subsume all phases from "seed to shelf." Where in 1900 the average U.S. wheat farmer got 70 percent of the cost of bread, today he or she gets only 5 to 10 percent.

The standardization of crops originated to a great degree because of the logic of commodity traders. Traders wanted a system whereby a bushel of wheat equals a bushel of wheat equals a bushel of wheat at any time anywhere in the world. This juggernaut led to a throttling of diversity, since traders certainly didn't want to cope with 300 varieties of wheat. Standardization of commodities facilitated trade, in contrast to the decentralized, differentiated state of the agricultural world at the turn of the century.

Ecosystem function is actually the most valuable aspect of the land, yet it has been ignored in economic terms. A recent valuation study of a Swiss forest, for example, found its greatest value, 40 percent, was from its ecosystem function, its abilities to hold soil, attract and hold moisture, produce oxygen, and provide habitat for biodiversity. A mere 5 percent of its value came from timber, and an additional 15 percent from recreation. In fact, as biodiversity prospecting has spread across the Third World by companies seeking valuable genetic materials, the value of biological materials alone is projected to run into the billions of dollars.

Consequently, any producer now attempting to use sustainable methods must absorb the costs of these practices, and then can't compete financially with the artificially low costs afforded to conventional ways of doing business. As Clay notes, no country in the world today has economic incentives for producers who avoid soil erosion or maintain long-term soil fertility. Those who mine the soil for short-term gain to produce cheap grain compel other producers to do the same or perish in the marketplace. The real costs are being passed on to future generations.

The fact that commodities are dominated by giant companies exacerbates the situation. In agriculture, for instance, a mere four companies own and control at least 40 percent of every major commodity—gargantuan oligopolies against which smaller producers have no chance of competing. Moreover, the traders themselves have absolutely no accountability.

SOLVING THE COMMODITIES CONUNDRUM

Clay sees several solutions to these problems. First, there must be a closer link between producers and consumers. Consumers seldom know the source of the goods they buy, much less the manner in which they were produced. Consumers are willing to pay for products produced sustainably. Green marketing and environmental premiums are viable mechanisms to pass the costs of sustainable production on to the consumer.

Second, the "rent" on goods must be adjusted to include negative environmental impacts. These environmental taxes will reflect a more realistic price on consumption. Consumers can then vote with their pocketbooks in encouraging sustainable resource use. A small one-cent tax per pound on sugar consumption would generate $250 million a year in the United States alone, funds which could be used to start cleaning up the Florida Everglades, which are being devastated by runoff pollution from huge chemical usage by the sugar industry. For context, Clay observes that the U.S. Agency for International Development spends a mere $150 million a year on the entire environment.

The third key solution is altering public policy to mandate the integration of environmental costs into the true costs of goods. Some analysts have calculated the true cost of a hamburger at 35 dollars, once the costs of soil depletion, water pollution, and biodiversity loss are factored in. National tax incentives can redirect large flows of capital into more sustainable enterprises.

Clay also points out that sustainable systems of production are labor-intensive. Consequently, reversing the trend of machine-intensive production, which displaces workers, will create large numbers of jobs, the first priority of a healthy economy.

These are the intentions behind Clay's latest ingenious venture. The support of worker-owned sugar plantations in Belize is helping to finance a worker buyout through an ESOP, an Employee Stock Ownership Program. An ESOP is a vehicle that permits workers to buy or own a large stake in a company, and ESOPs have been successfully used by many large companies, including Avis, United, and America West. By making the workers stakeholders, an ESOP gives a high incentive to employees to succeed. The periodic failure of ESOPs is mainly rooted in the fact that workers often don't know how to manage a company. Consequently, Clay's scenario mandates that outside professionals, in providing the capital for the buyout, take a strong hand in management, at least at the outset. Then a systematic transition toward sustainable production methods will be implemented, adding value and differentiating the product in positive environmental and social terms. Many companies have shown themselves willing to pay premiums for raw materials from producers using more sustainable practices and sharing the benefits with greater social equity. Clay has identified sources of international capital favorable to furthering such a transition, and he is heading south with these commitments.

OFFERING OPPORTUNITIES

Clay traces his affinity for economics and ecology to his own upbringing on a "Norman Rockwell" farm in Missouri. The 156-acre farm, which had no indoor plumbing or phone till he was 15, grew mixed corn and soy-beans, and raised cattle, pigs and chickens. He hunted rabbits and deer and fished for about half the family's meat diet, and the family had large subsistence gardens and orchards.

"I spent a lot of time outdoors," Clay recalls. "I planted, but I also camped out. I knew all the species, all the trees by bark, seed, and twig. I learned the different relationships of species to one another. I gathered nuts, berries, and mushrooms. This knowledge was important in my family. When I was eight years old, I remember having to choose which and how many rabbits to kill that day: If I shoot this rabbit, are we going to have rabbits in three months? If it was a larger, probably female rabbit, I wouldn't shoot her, because you know you need to think about the future. A lot of people close to the Earth think about that all the time—not consciously—it's just the way they live."

Clay's father died in a tractor accident when the boy was 15, and he helped his mother run the farm for four years before going off to Harvard on scholarship. "Much of what I have done with my life is to try to offer people opportunities, not handouts," he reflects. "I've had a lot of opportunities and believe other people should, too."

At college, Clay planned to study law and go into politics, but his interest shifted to anthropology and the social aspects of economics. After attending Harvard and serving in Vietnam, he began to study traditional cultures in Mexico and Brazil. He completed a Ph.D. at Cornell on international development, global agriculture, and anthropology, and then started teaching at Harvard. After a stint with the U.S. Department of Agriculture, seeing the negative impact of development aid overseas, he signed on with Cultural Survival in 1980.

"For the last twenty-five years I have been trying to figure out a way at a much larger level to get back to the kind of system that I grew up with, but a much more balanced system. Subsistence was a huge chunk of our farm. Self-reliance was a huge chunk. Getting back more toward those models is something I've been working at. For me personally, a lot of the importance is economic balance. Part of it is finding pleasure in a direct relationship with the Earth. Part of it is an appreciation of the diversity in the world. Social justice is the balance of how we walk on the Earth."

175

Indeed, as Clay packs his bags to travel to this most unorthodox buy-out of a plantation, the possibilities for environmental sustainability and social justice seem considerably brighter for it.

"In Brazil," he relates, "it is a custom that a young girl plants palm trees that will bear fruit at the time she has children and needs the fruits to feed her own children, just as her mother did for her.

"The Oromo nation in Ethiopia has a proverb: 'You can't wake a person who is pretending to sleep.' We have to wake up and look at the system we've created. If you can take the kaleidoscope and shift it just one notch, there are all kinds of possibilities out there."

CHAPTER SEVEN

Redesigning Society:
The Second Industrial Revolution

IN CARTOONS, WE OFTEN SEE A LIGHT BULB SWITCHING ON TO SYMBOLIZE that a character is getting a brilliant idea. Actually a light bulb is only 4 percent efficient, wasting 96 percent of the energy it uses. A light bulb is a space heater that gives off a little illumination, according to energy expert Amory Lovins. Lovins goes on to note that cars use only 2 percent of their energy to move and stop the wheels! In fact, it appears that our vaunted industrial system is 94 percent inefficient, wasting over nine-tenths of a product before it even gets to you. We call this an effective system! Perhaps the light bulb ought to symbolize dim ideas, or what populist political humorist Jim Hightower likes to call a "five-watt bulb in a hundred-watt socket."

The good news is that sustainable industrial practices and "green design" approaches do exist and already function effectively in many locales. While they still represent only a tiny fraction of current economic activity, they are viable and inspiring models. Because they are far more efficient and economical, they are gaining serious attention in the business community because they could represent both competitive and marketing advantages. In retrospect, our current practices will likely look like the Dim Ages of the five-watt bulb.

THE PROBLEM

Our industrial systems are voracious consumers of finite natural resources, which they then spit out in largely indigestible form as waste. Our global economy is based on the gargantuan use of synthetic chemicals whose ecological impact and negative health effects are clearly implicated in a wide range of environmental and health problems. While industry has succeeded in supplying remarkable levels of material goods to large numbers of "consumers" in developed nations, it is moored to the illusory dock of infinite resource availability. This fantasy of limitless growth has exacerbated social inequities and exploited resources from the poorest, most powerless regions and people, while concentrating environmental and social devastation there. The system wastes people as well as stuff.

The amounts of resources our industry uses are almost unimaginable. The average U.S. citizen consumes about 25 pounds of basic materials a day (at least). "Daily consumption at these levels translates into global impacts that rank with the forces of nature. In 1990, mines scouring the crust of the Earth to supply the consumer class moved more soil and rock than did all the world's rivers combined. Since 1940, Americans alone have used up as large a share of the Earth's mineral resources as did everyone before them combined."[1]

The waste we generate is also astonishing. The average U.S. citizen produces three-quarters of a ton of garbage annually. According to the Worldwatch Institute, four cents of every dollar Americans spend on goods goes to packaging. In industrial countries, packaging makes up close to half the volume of municipal solid waste. About 200 billion bottles, cans, plastic cartons, and paper and plastic cups are made and thrown away yearly.[2]

The death of forests is one potent indicator of industrial harm. "Petrochemical smog" from pesticides, fertilizers, and vehicle exhaust gases returns to kill forests as acid rain, heavy metals, nitrogen oxides, and ozone. As early as the 1960s, ozone diminished growth rates by 80 percent in the San Bernadino Forest south of Los Angeles. By 1991 in the United Kingdom, 56.7 percent of trees were at least moderately defoliated.[3] (There is currently also some improvement in Europe, as certain types of air pollution have started falling.)

Radioactive waste generated by the nuclear energy industry is currently an unsolved problem. Yet our most mundane wastes may prove as overwhelming as these most lethal ones. Incinerators release heavy metals into the air and produce large quantities of fly ash filled with toxins, which is then dumped in a landfill. Many landfills are rapidly filling and leaking into groundwater. Waste disposal costs a great deal, and taxpayers bear the costs of this pollution. Many estimates for cleaning up hazardous-waste sites in the United States range toward $750 billion. But the core issue is that so far none of the conventional "solutions" really "clean" toxic waste, but rather constitute a shell game of moving poisons around, hiding them, and changing their shape but not their danger.

The planetary nature of economic activity permits the main players — the transnational corporations — to hide environmental devastation in the poorest countries and regions, while providing the illusion of a never-ending material cornucopia to the world's privileged. According to Greenpeace, at least 10 million tons of waste of all types have been exported over the last several years from developed countries, more than half of which have gone to Eastern Europe or developing countries.

The global toxic economy is disrupting our planetary ecology, causing perturbations of natural systems that regulate our biosphere's climate and atmosphere in ways that we barely understand. Eight of the hottest years ever recorded have come since 1980. Ozone depletion in the upper atmosphere is worse than previously suspected, and the resulting increase in ultraviolet radiation will have serious effects on plant, animal, and human life, including the loss of much plankton, the base of the entire marine food chain and a critical supplier of oxygen to the global atmosphere.

The United States, with only 5 percent of the world's population, consumes roughly 40 percent of the world's produced resources. Meanwhile, fully half the world's people are undernourished, and one-fifth live in extreme poverty. The rest of the world, which aspires to our MTV lifestyle, simply cannot attain this level of prosperity without immense ecological harm. For example, feeding the world on the diet enjoyed by the average U.S. citizen based on the same level of inputs into agriculture would require all the world's current oil production and exhaust known reserves within a decade.[4] The developing world, with far less efficient and more polluting energy technology, is emulating our industrial model by increasing its automotive fleet and manufacturing sector.

When one adds inefficient heating and lighting and an agricultural system that ships the average bite of food 1,200 miles from farm to plate, it is clear that the industrial system is simply not sustainable.

SOLUTIONS

Vastly superior alternatives to our current industrial technologies are already gaining the attention of both large companies and the public policy makers of industrial nations. A number of very positive trends are emerging.

Increasingly, utilities have begun using some renewable energy sources, superefficient technologies, and vigorous conservation incentives. The increasing financial competitiveness of renewable energy sources—wind, solar, geothermal, and biomass—is spurring the growth of those industries. These four sources provided 11 percent of California's 1993 total electricity. According to the Worldwatch Institute, the potential for global wind energy alone is five times the current world electricity use, and all U.S. power consumption could be satisfied with solar plants. None of these resources is without drawbacks, but used intelligently, they would be far less destructive than our current energy sources.

One pioneer is California's Sacramento Municipal Utility District, which has replaced 42,000 energy-guzzling refrigerators and planted half a million shade trees. It purchases power from four industrial cogeneration plants, and has invested in a wind farm and electric cars. It is installing solar electric collectors on customers' rooftops. All these factors have lowered the cost of electricity. Other examples of successful alternative energy technology usage can be found around the world, although they still account for only 1 percent of global energy production.

It is also possible to increase the efficiency of much industrial production by 30 to 90 percent with new designs and adjustments. New, more efficient appliances (especially refrigerators), tungsten halogen and compact fluorescent bulbs, better home design, insulation, and more use of solar heating could all combine to reduce domestic energy use radically.

A few firms have led the way in pollution prevention. The "Pollution Prevention Pays" program of the large 3M company encourages employees to redesign products and processes to use fewer or no harmful chemicals. Estimates by 3M show prevention of more than a billion pounds of emissions and cost savings of $500 million since 1975. Following 3M's example, large companies including Dow Chemical and ICI have instituted comparable programs. Ecover, a leading European company with a strong environmental commitment, built a soap factory which has virtually no harmful emissions.[5] Intel, a major U.S. computer chip manufacturer, tweaked its manufacturing processes and process controls, and thereby diminished hazardous waste requiring treatment and disposal by 95 percent, while its annual sales rose 500 percent.[6]

In the area of reducing waste and recycling, Germany has led the industrial world with its ambitious DSD (Duales System Deutschland) mandatory manufacturer-financed recycling program. BMW is beginning to design cars with disassembly in mind, with an ultimate goal of 100 percent reusability. In Japan, manufacturers are being required to label parts for recycling and to open resource recovery centers for old machines. New Matsushita washing machines can be completely disassembled with a screwdriver.[7]

Mandates for increased auto fuel efficiency and eventual conversion to electric, solar, or hydrogen fuel-cell auto engines would begin to rein in the damage of the automobile age. The "hypercars" that Amory Lovins has boldly proposed to car manufacturers hold the potential to revolutionize the industry with cars that can travel across the U.S. on a tank of fuel with almost no pollution. Many quality-of-life questions still remain regarding how much blacktop people and the planet need and want.

A partial solution is to transform manufacturing to "industrial ecology." According to the World Resources Institute, "One place where this challenge has been met, albeit on a limited scale, is the town of Kalundborg, Denmark. Here, industrial waste and waste process heat are exchanged in a cooperative arrangement among a power plant, an oil refinery, a pharmaceutical manufacturer, a plasterboard factory, a cement producer, farmers, and the utility that provides residential heat for local residents. The arrangement, which has been financially beneficial to all parties, is a working model of a small industrial ecosystem."[8] Systems such as these still rely on toxic materials, however.

WHAT YOU CAN DO

Clearly many of the solutions to problems resulting from poor industrial design must be implemented on a societal scale, so getting involved politically is appropriate both in your local community and in national affairs. Several groups listed in the Resource Section can help direct your efforts. But at all stages you can minimize your usage of toxic and non-recyclable materials. You can also try to conduct more of your economics in your neighborhood and region, buying locally produced organic foods, using a bicycle or your feet whenever possible, and evaluating the resources that your lifestyle demands to see whether renewable resources can be substituted. Perhaps you can put a solar collector in your home to heat your hot water. Growing some of your own food also helps. Start a compost pile or a neighborhood compost operation. Supporting companies working in these areas also makes a difference, and these companies can often help you learn more about what to do about the absurd waste our society engenders.

The technologies, the designers, and the models exist, but we must also be willing to rethink our lives in a global effort to redesign our society within natural limits, as Monika Griefahn has done in Lower Saxony in Germany described in this chapter. Countries like Germany, Denmark, and Holland have taken meaningful steps to institute the Next Industrial Revolution to create vast new industries without harming the Earth and wasting people and resources as we do now.

The real lasting solution to our crude way of life today is to emulate the waste-free food loops of ecosystems. Before building anything, or using a material or chemical, this new outlook calls for a Life Cycle Assessment (LCA), thinking through all the implications from resource extraction to disposal, measuring the real environmental and social costs. William McDonough demonstrates that natural substitutes can be found for toxic building and industrial materials and systems. McDonough is now working with large manufacturers to redesign their industrial loops to eliminate waste and substitute nontoxic natural products, which can be thrown in the garden for compost after usage.

Monika Griefahn and the New Ecological Deal

Monika Griefahn is translating the lessons of ecologically based economics into tangible political action in society. She is implementing environmental restoration through governmental channels by taxing environmentally harmful practices and supporting environmental technologies while creating jobs, improving quality of life, and midwifing a lucrative new industry for Germany. She is cooperating with industry to create an economic model of sustainability that will receive worldwide attention.

EDWARD ABBEY, SELF-STYLED ANARCHIST AND INSPIRATION TO A GENERATION of environmental activists, once commented that the major problem today is popular disenfranchisement by elitist power politics. Abbey suggested that things wouldn't be politically right with the world until we are all gathered around campfires again with painted faces to make the political decisions affecting our lives. His vision may have been a touch more poetic than literal, but at the core what Abbey was legitimately invoking was the lost sense of democratic decision making on the ground level. A fierce individualist, he nevertheless recognized politics as the expression of community, a group will elevated to collective action.

While as citizens we can make meaningful strides toward improving the environment by direct individual action, other changes will occur only at the level of public policy. They must involve the nations and governments of the world.

But what would a green society look like—a vision of a positive environmental future? When the Turner Foundation put out a call to fund a novel based on such a vision, it was unable to award the grant because of the lack of a winning entry. This disturbing absence of vision underscores the dilemma that it's hard to get somewhere if you don't know where you want to go.

183

THE NEW ECOLOGICAL DEAL

A positive vision is indeed emerging from Europe, a land mass about the size of the United States supporting 350 million people, and a classic example of how ecology does not heed political borders. The European Union is acutely aware that it must collectively deal with environmental misbehavior as polluted water tables migrate around the continent, pesticides drift in the wind across national boundaries, and air pollution blows from one country to another.

In Germany, an important model is illuminating the beginning of a green path for society. Leading the effort is Monika Griefahn, activist turned politician, in a country that has birthed the most advanced green political movement in the world today.

"To overcome the lack of economic perspectives as well as ecological instability," says Griefahn, tapping a pen on her unassuming desk in Hannover, where she presides as Minister of the Environment for the state of Lower Saxony, "we support the initiation of a reform process which aims at an ecologically oriented change of structures to modernize the economy of society. Following the great program of social reforms introduced by U.S. President Franklin D. Roosevelt to overcome the worldwide depression in the thirties, we suggest an 'ecological deal.' As in Roosevelt's 'New Deal,' we have to make a new start instead of getting deeper and deeper into a situation from which there is no way out."

The changes to which Griefahn is alluding are neither trivial nor academic. They incorporate the historical German penchant for efficiency, and they also hold the potential for alleviating the country's current retrograde economic status caused by the unification between East and West, which has stalled what was formerly among the world's strongest modern economies.

If Griefahn is right, markets for environmental technologies could prove a growth engine for Germany in the foreseeable future. "Only an economy which switches over to environmentally friendly products and processes secures the basis for its own existence in the long term," says the auburn-haired official cheerfully, seated casually beneath 14-foot carved baronial ceilings. "One of the primary aims of the government of Lower Saxony is the ecological reorganization of industry."

ENVIRONMENTAL SIN TAXES
AND INDUSTRIAL REHABILITATION

The Ecological Deal is founded on a state investment program financed by taxes from waste disposal and water extraction. It is based on ecological infrastructure investments and the environmentally sound restructuring of industry. Griefahn points out that the investment in Germany for environmental technologies is today well over 40 billion deutshemarks (dm), or $25 billion, while the number of firms operating in the market has quadrupled to over four thousand in just ten years. She would like to see the number reach 600 billion dm ($375 billion) to effect massive change. In fact, Griefahn adds, "The responsibility for the protection of the environment has to be transferred more than at present to companies, and not be borne by the public, as is still often taken for granted."

In practical terms, this strategy entails redesigning industrial society from cradle to cradle, as architect William McDonough describes (see the next section in this chapter). McDonough's business partner, Michael Braungart, who is Griefahn's husband, is a leading-edge industrial chemist who has undertaken the redesign of the entire flow-based industrial cycle using natural and sustainable materials. The system focuses a priori on avoiding waste, and on holding the industrial producer to ultimate responsibility for the return and recycling of materials. The program instituted by Griefahn's Social Democratic Party (SPD) is part of a Commission on the Protection of Humanity and the Environment aimed at testing ecological flow management and translating it into legislation. The analysis also mandates the inclusion of "external costs," that is, environmental costs ordinarily ignored by current blind-side economics, in the cost of the product.

The program lays outs seven principal areas of corrective activity, providing incentive by government tax breaks and subsidy programs, using public policy as a staff to shepherd the recalcitrant industrial flock. First, according to the legislation, these "intelligent products" and production techniques must not contain any substances or compounds that accumulate in living things and are not biodegradable. The legislation bans the use of known carcinogens and mutagens. It calls for the least usage possible of harmful substances. By assigning the ultimate product liability to the manufacturer, accountability is hard-wired into the system where it counts — in corporate profits and liability.

This methodology is already being incorporated into German manufacturing processes. Cars built by VW, which has its headquarters and ten thousand workers in Griefahn's state of Lower Saxony, are now made from entirely recyclable or reusable materials, and have reduced materials usage altogether. All defunct cars have a guarantee to be returned to the manufacturer, which dissembles the vehicles and reuses many of the components. VW is working to eliminate as many toxic or harmful materials from the cycle as possible. The practice is proving to be economical as well as ecological. VW has also developed gas-saving engines that get almost sixty miles to the gallon.

Economic policies already in Lower Saxony reflect this new way of looking at the true costs of goods. The state instituted a special waste tax which brings about $10 million a year into the fund to develop waste-saving and waste-avoiding technologies and practices, as well as to create a consulting agency for waste reduction and avoidance.

Similarly, the state levies a water tax to discourage the wasteful use of water and to invest in reclamation programs. The private and industrial use of water will become more expensive, Griefahn notes. Water pollution is endemic in a land where 80 percent of drinking water comes from wells tapped into groundwater. Groundwater pollution is now severe and widespread, and Europe's rivers are in notoriously bad shape, though undergoing slight improvement. Water consumption must be reflected in appropriate cost accounting, says Griefahn, and the $90 million annual proceeds from the levy applied toward the protection of natural water resources. "We also pay for river regeneration programs," says the minister, "so that in the future we can drink from the rivers again. It's a dream — you need wishes."

Griefahn's public policy underscores the basic tenet that industry has been receiving a free lunch at the public trough. The real environmental costs of energy and raw material extraction must now be factored into the cost of goods. Recognizing the inevitable logic of the system, industry in Lower Saxony has expressed its willingness to cooperate, and participated with the government in a "waste summit" in Hannover in 1992, agreeing to reduce the output of toxic waste by 40 percent over the next five years.

Griefahn's next main agenda is the predictable increase of fossil fuel prices to reflect their true environmental costs. This program will divert fossil-fuel tax revenues to fund an investment program for climate protection, energy efficiency, energy conservation, and another program to substitute renewable energy sources as the main sources of energy by the year 2050. The plan will phase out coal and nuclear in the medium term to replace them with solar, wind, and other appropriate energy-conservation technologies.

COMPANIES SIGN ON

Even large companies are coming forward in support of the ecological tax. Griefahn cites Michael Otto Distribution Centers, a large retail operation with sales around 23 billion dm ($14 billion), which has said to her, "If you want us to do the right investments, we need an incentive, and that's why we need ecological tax reform." The large companies have complained to her that they are compelled by the financial system to invest in machines that put people into unemployment instead of putting investments into machines that unemploy energy or resources used. She says the new companies working with solar energy, wind generation, and compostable plastics are performing well financially by benefiting from such incentives, and they are seizing a leading market position at the same time.

In Lower Saxony, the ecological taxes are helping finance a sustainable future. About $150 million have been generated in an Ecology Fund to seed small and medium-sized companies or communities devising ecological technologies. Jointly managed by the Ministry of Economic Affairs and the Ministry of the Environment, the fund has received well over 7,000 applications, of which it accepted 6,200. The fund has in turn generated 400 million dm ($250 million) of additional investment from the private sector. Among the projects funded are energy conservation, new and renewable energy technologies, waste-avoidance technologies, environmentally friendly products, environmental education, and environmentally responsible tourism.

INCUBATING A POSITIVE FUTURE

Griefahn points out that a study conducted by her government shows that 30 percent more light trains can run on existing tracks. Given the fact that the transport of goods in Europe is projected to double in the next ten years, the cost savings and environmental improvement represent potentially dramatic gains. She further underscores the importance of wind energy, a clean, safe, and renewable source which her government has also studied through the Ecology Fund. "I think we now have more efficient wind generators than anybody in the world." The Minister of the Environment beams proudly. "We created a program that allowed people to buy wind generators and enhanced their performance. They now generate three times more energy than they did three years ago, which makes them cost-effective. So now it becomes a market, and we want to make it an export market."

Griefahn has also used the buying power of the state of Lower Saxony to effect environmental improvement. The state now uses not only economic criteria but also ecological criteria for its purchases. Through a public sector procurement program, the government is awarding building contracts and orders for goods and services to ecological suppliers. "When all social costs are taken into consideration," says Griefahn carefully, "the most ecological solution is generally also the cheapest." She points out that the annual procurement demand at the federal, state, and regional levels in the Federal Republic of Germany is 80 billion marks ($50 billion), money that now carries the weight of policy.

Because of the financial incentives to industry to clean up its act, Lower Saxony has also carried out a pilot experiment on ecological comptrolling with three medium-sized companies. A step beyond an environmental audit, which mainly helps a company conform to self-induced standards, the comptrolling process examines the permanent cycles in which all the effects of a company's impact on the environment are continuously recorded, assessed, and redirected according to environmental and economic efficiency goals.

Griefahn deliberately chose companies with open-minded management whose products are linked with special environmental problems, including makers of foil packaging and suppliers to the paint and varnish industry. According to Griefahn, "The pilot experiment has shown that environmental comptrolling was able to significantly improve company orientation in the production sector, especially the level of information on ecological conditions relevant to the company. In recognizing their interdependence with the environment, the companies have been able to convert their production to more environmentally friendly processes and to develop measures to conserve raw materials, energy, and water." One company saved large amounts of energy and money by constructing a cogeneration power plant utilizing "waste" heat.

The experiment led Griefahn and the SPD to conclude that environmental comptrolling needs to become embedded in the companies' day-to-day decision making. The state is now preparing business handbooks to help businesses seeking to transition to more ecological functioning. Lower Saxony is working closely with the furniture and carpet industries. The Ecology Fund is prepared to provide grants to cover 30 percent of the costs of integration for companies ready for environmental comptrolling.

Another tenet of the SPD legislation is to alter public transport policies. Undoing the nightmarish traffic that has become a clot throughout Europe is a social as well as an environmental priority, and can be achieved through greater reliance on advanced railroad technologies and bicycles. The goal to reduce petroleum consumption in half by 2005 is eminently doable, according to the SPD analysts, and truck pollution can expeditiously be reduced by the simple substitution of natural gas as fuel. Cars are to be banned from inner cities, a quality-of-life issue for many auto-exhausted urbanites.

Griefahn adds that we must make a critical societal distinction between living standard and quality of life. The high living standard of industrial society is also plagued with the precise absence of quality of life. She cites the new model she herself enjoys in Hannover. "For me, it's very attractive to buy in my neighborhood, not to drive to the supermarket to go shopping, or to need a car to drive my kid to the doctor. It is attractive to have things in your neighborhood without having to go anywhere by car. You cannot change the cities, but you can restructure the way by which you have connections again between your neighborhood and your daily needs, such as work and food. Close to my house, for instance, is the school where once a week a farmer comes and sells his products of ecological farming. I do all my shopping there because I can go there on Saturday morning and get good food and everything I need."

This decentralized model is dramatically opposed to the current European model, which centralizes food production and thus entails long distance transport for distribution. Griefahn again emphasizes the need for ecological tax reform to build in the true costs of such inefficient energy-intensive methods. At the top of her list is the transformation of agriculture, which can be accomplished through the linkage of grants and subsidies with ecological criteria. Her office walls are graced with striking photographs of indigenous grain varieties.

GREEN POLITICS

The Ecological Deal also calls for the industrialized countries to take responsibility for closing the ozone hole. Griefahn is pushing for immediate legislative action to stop the production and use of fluorochlorohydrocarbons and certain compounds, such as the pesticide methyl bromide, which directly contribute to ozone damage.

For Griefahn, pollution and environmental hazards are not an abstraction. Growing up in the Ruhr valley, a coal and steel region comparable to sooty, industrial Manchester in England, she remembers "seventeen million people on a very little spot and no green in between the towns." Surrounded by streaming smokestacks, she became sick and was sent to recover each year at the seaside to stop her coughing. "I thought that shouldn't be okay. It should be that wherever you live, you can live healthily. It was not conscious in my mind then, but later as a college student, I began organizing citizen groups against environmental pollution and nuclear power."

While conducting environmental education campaigns, Griefahn came in contact with Greenpeace, newly formed in Europe in the late 1970s. Inspired by the "Big Green's" direct-action campaigns and no-nonsense approach to solutions, she founded in 1980 the German chapter of Greenpeace, which grew to be the biggest in Europe. She conducted campaigns against chemical pollution of the North Sea and dumping of radioactive wastes in the North Sea and Atlantic. In 1983 she was elected to the Greenpeace International Board, the only woman on the board from 1984 to 1990. Griefahn took the German branch from zero to 700,000 members, among the largest and most effective in the expansive global Greenpeace network. She notes that, according to polls, the German people give the organization a 72 percent acceptance rating, in contrast to a high around 20 percent for politicians.

Griefahn hardly expected to become one of the very politicians she spent much of her time exposing and pressuring. After Gerhard Schroder, also a former Greenpeace activist, won the elective office of prime minister of Lower Saxony in 1990, he insisted on appointing her Minister of the Environment for Lower Saxony. "For me it was good to say, 'Okay, for ten years I have demanded of politicians what they should do, and now I will try to do it myself.' What I can do to change things is from the state level, not so much from the federal or European level. But it's a starting point."

Lower Saxony is a geographically large state with industry such as Bosch and VW and a population around 7.2 million, making it the third or fourth largest in the country. It is probably the most progressive state, and from it emerged a powerful and seminal coalition between the German Green Party and the larger Social Democratic Party.

In Germany, the Green Party has evolved through a dynamic sequence of incarnations, beginning as an opposition party that had no interest in gaining electoral office. However, when the Greens—or *Die Gruenen*—gained more than the necessary 5 percent of the popular vote to warrant seats in parliament allotted proportionally according to vote shares, they were compelled to make new choices about how to be effective. After politically disastrous adventures with foreign policy issues, the Greens reverted to a tight focus on two principal issues: the environment, their original cornerstone, and women's rights.

In the last elections in Lower Saxony in 1994, the Greens got a healthy 7 percent of the vote, a gain of 1.5 percent. In four years of the Red/Green coalition between the Greens and the Social Democratic Party, together the two have gained by over 3 percent. "We have won by the effects of what we done," Griefahn surmises, "by the programs we have done. People are convinced, and so we have won. It is respectable."

191

Griefahn foresees that environmental protection will emerge as a national objective codified in the constitution. She has also been appointed head of the national Committee on Environment, Nature Protection, and Nuclear Issues in the German federal Bundestag, and she is actively working on the national level for countrywide change. She is also calling for an environmental union of the European nations along with the current economic union.

But, Griefahn says, to resolve environmental problems, there must be larger global political alliances among nations and states in acknowledgment of the whole Earth as a living ecology. To her delight, the city of Hannover will have the opportunity to play a major role in highlighting the passage to a green future.

A WORLD'S FAIR OF SOLUTIONS

Hannover's only fame has been as an a showplace for international trade fairs. The city, which often hosts industrial, computer, and other commercial trade shows, is now gearing up for Expo 2000, the World's Fair for the millennium, which will be devoted entirely to sustainability. Griefahn intends to show the world that the solutions to our environmental problems are present. "Worldwide efforts require common aims, which must be implemented as regional solutions. The global communication of information has to be improved to disseminate such developments.

"We invite nations to bring and show solutions," Griefahn explains, "but we want to produce the Expo in a way that is not something like Disneyland, or that is destroyed afterwards. We will build fair halls that need no artificial climatization, and are energy-efficient. We will bring 90 percent of the people by public transport instead of cars. We will show how you can feed 600,000 people in an ecological way. One big theme is agriculture and the world of tomorrow, which will demonstrate sustainable food production combining the best of organic agriculture and traditional farming."

The theme of the Expo is "Humanity—Nature—Technology," and Griefahn is seeking to create an awareness that "an enduring coexistence between humanity and nature can only be maintained if there is a change in the use of technology." She sees it as "an important contribution to the ecological transformation of industrial production."

Griefahn also sees Expo 2000 as a crucial bridge between the industrialized nations of the North and the lesser developed countries of the South. She is calling upon the North to "outline perspectives to reduce their excessive consumption of resources to a sustainable level," and to cease their exploitation of Third World countries by changing their attitudes. She points out that the North, 20 percent of the world's population, consumes 80 percent of the world's resources, much of them from the Third World. In turn she calls upon Third World nations to "offer solutions to achieve a progressive improvement in living conditions with adapted technologies." This strategy includes population control.

As Griefahn assiduously implements her working model for industrial transformation on the regional level, she is patiently stretching toward a global vision. "I have not been a professional politician. I was not a member of a party when I was elected minister of the environment. I really want to influence economics, business, and finance. I come with a vision of where the world should be in the next millennium, and that's what I'm working for. I have little children, three and six years old. We can't wait for them to change the world; we have to do it now."

If Edward Abbey were still around, he'd be building a campfire outside the Ministry of the Environment and inviting Griefahn over. No doubt she'd bring the nontoxic face paints.

William McDonough and
the Next Industrial Revolution

*William McDonough is a visionary architect who is revolutionizing our very
concept of design structure. Working with ecological principles, he is bringing
architecture into balance with the natural world. He has also undertaken
a systematic assessment of existing toxic and unsustainable materials and
commenced a coherent substitution process using safe organic or industrial
materials.*

LEANING INTENTLY ACROSS THE OFFICE TABLE AGAINST THE HIGH-RISE
backdrop of New York City, William McDonough is wryly recalling Gregory
Bateson's ecological parable from New College in Oxford, England. The
stately beams of the university's main hall—timbers 40 feet long and 2 feet
thick—were finally caving from dry rot. The replacement cost would be
around $50 million, or $250,000 a log, not that anyone in 1985 could imagine where to find such a priceless wonder as a straight 40-foot English oak
from an ancient forest. Someone suggested querying the Oxford University
forester as to whether any of Oxford's own lands might still harbor such
trees. The forester had been wondering if anybody would ever ask. He was
relieved to remind the group that when the main hall was built 350 years
ago, the architects specified that a grove of oaks be planted and maintained
to replace the beams when dry rot would set in three and a half centuries
later. Of course, many generations of foresters had taken care to assure that
the trees would be there on time for the latter twentieth century.

Surveying the ecological devastation that is Manhattan, where civic
leaders did not plan for the sustainable replacement of New York City's
infrastructure, McDonough sits at the fifteenth-floor glass bow of his
25-person office. A miniature village of neat, white cardboard scale models of dozens of projects he is designing stretches the length of the capa-
cious rooms behind him, as draftspeople scurry about like giants in a
miniature world.

194

"WHO IS THE LEADER OF THE SHIP?"

The inspired green architect launches into a typical star burst of Big Ideas. "Peter Senge, a professor at MIT's Sloan School of Management, has a learning laboratory, part of which is a leadership program. One of the first questions he asks CEOs of companies is, 'Who is the leader on a ship crossing the ocean?' He gets obvious answers such as the captain, the navigator, or the helmsman. But the deeper answer is the designer of the ship, because the success of operations on a ship is a consequence of design, which is the result of human intention. Today we are still designing steamships, machines powered by fossil fuels that have deleterious effects. We need a new design.

"If we understand that design leads to the manifestation of human intention, and if what we make with our hands is to be sacred and honor the Earth that gives us life, then the things we make must not only rise from the ground but return to it—soil to soil, water to water—so everything that is received from the Earth can be freely given back without causing harm to any living system. This is ecology. This is good design."

Nattily dressed with wide gray suspenders and a blue and gray bow tie, McDonough could appear charmingly archaic, but in fact his revolutionary ideas about design have challenged the modern canons that underlie the structures not only of buildings, but of society itself. The byword is *sustainability*. "Sustainability should mean meeting the needs of the present without compromising the ability of future generations of all living things to meet their own needs," he states precisely. Or, as Thomas Jefferson might have rephrased it, "To enjoy life, liberty, and the pursuit of happiness free from intergenerational remote tyranny," he adds.

"Our culture has adopted a design stratagem that says, 'If brute force doesn't work, you're not using enough of it.' The classic architectural challenge has always been how to combine light with mass, membrane, and air. Contemporary architecture, which arrived with the advent of cheap glass and cheap energy, allowed architects to stop relying on the sun for heat or illumination. We made glass buildings that are more about buildings than people. The hope that glass would connect us to the outdoors was completely stultified by making the buildings sealed. People are trapped instead, and indoor air quality issues are now looming treacherously with the thousands of chemicals used to make things today."

McDonough points out that a patriarch of modern architecture, Le Corbusier, was proud to call his buildings "machines for living in." The mechanistic fetishism of the early twentieth century celebrated steamships, grain elevators, airplanes, and smokestacks as signs of "progress." McDonough asks rhetorically whether an office is a machine for working in, or a cathedral a machine for praying in? "Designers are designing for the machine, not for people. We need to listen to biologist John Todd's idea of 'living machines,' not machines for living in."

PRINCIPLES OF NATURAL DESIGN

McDonough finds three basic laws of the natural world on which to model human designs. The first principle of natural design is that nature does not have waste. In nature, waste equals food. "It's not a question of reducing waste," he says decisively, "but of eliminating the concept of waste. Therefore everything becomes a *product*. Everything is being cycled constantly."

The second principle is energy, the natural force that animates life and its cycles. "This energy is the only thing that comes from outside the terrestrial system, and it's in the form of *perpetual solar income*. Not only does nature operate on 'current income,' it does not mine or extract energy from the past, it does not use its capital reserves, and it does not borrow from the future. It is an extraordinarily complex and efficient system for creating and cycling nutrients. If we look at the use of fossil fuels and mining from a designer's perspective, you can see that a detoxification happened on this planet over millions of years by rendering the surface of the Earth habitable to people because a lot of these toxins were buried. Now we are mining it at a furious rate. We're retoxifying."

The third element of natural design is biodiversity and diversity of place. "What prevents living systems from running down is a miraculously intricate and symbiotic relationship among millions of organisms, no two alike. That means that cultures everywhere start to develop their own architecture, working with indigenous materials that don't come from distant places, that relate to diurnal and annual cycles."

196

McDonough offers the classic examples of the Bedouin tent and adobe home. In the Arabian desert where temperatures routinely soar to 120 degrees without shade or breeze, the black Bedouin tent creates a deep shade that lowers temperature to 95 degrees. Its coarse weave makes for a beautiful interior illuminated with a web of lights. Because of the coarse weave and black color, the air inside rises and passes through the membrane, causing a breeze to enter from outside and cooling it further. When it rains, the coarse fibers swell, and the tent becomes watertight.

Similarly, classic adobe mud-brick structures create a thermal mass to store a winter's day sun's energy for the cold night. Conversely, adobe reflects the heat of the summer sun and stores the coolness of the evening. But McDonough is not suggesting going back to nomadic tents and mud huts. To the contrary, his innovative work has combined high touch with high tech, customizing ancient folkways with modern technological advantages.

A GREEN INDUSTRIAL REVOLUTION

When the American Institute of Architects discussed creating a "green award" for design to honor McDonough's kind of work, he protested that neither he nor anyone deserved it. He stated that the industry was not meeting minimum standards of sustainable design. The architect had become painfully aware of the existing limitations to meeting such criteria when he was contracted to design the first of the green office projects, an ecological, nontoxic New York headquarters for the nonprofit Environmental Defense Fund (EDF). "I can't make the buildings I want to make," he found. "Once you get past the frustration of realizing that the materials we're encountering every day are essentially toxic, after you get past the depression, you wonder how to do it another way. What I'm saying is that it's not 'reduce, reuse, recycle.' It's 'redesign, reduce, reuse, recycle.' I'm looking at how to make a *new industrial revolution*."

In search of what proved to be nonexistent nontoxic building materials for all his designs, McDonough linked with European industrial chemist Michael Braungart, who was working extensively redesigning basic industrial components themselves. The inventive chemist delineated three typologies for products: *products of consumption; products of service; and un-marketables.* To achieve sustainability, all these components must be considered as products, since nature does not have waste. McDonough also joined forces with innovative economic theorist and author Paul Hawken, who considers him one of the leading architects of industrial design, and which he wrote about in *The Ecology of Commerce.*

"Consumables," says McDonough empirically, "are products that when eaten, used or thrown away literally turn back into dirt, and are therefore compatible with other living organisms. They could be returned to the ground to restore the life, health, and fertility of the soil. Shampoos could be in bottles made of biodegradable plastics to go in your compost pile instead of the landfill. Furniture made of lignin, potato peels, and technical enzymes can be safely returned to the Earth." Today a shocking 30 percent of landfills are choking on construction debris, and an estimated 54 percent of U.S. energy used is related to the construction industry.

What we have called "durables," such as cars or TV sets, are actually products of service, because customers want the service, not the product as such. To eliminate waste, such products should be licensed or leased rather than sold. "I've got this item here," says McDonough with a Strangelovean gleam in his eye. "I'm not going to tell you what it does, but you're going to love it. Before I tell you what it does, let me tell you what it is, and you tell me if you want it in your house. It's four thousand and sixty chemicals, many of which are toxic and two hundred of which off-gas into the room when it is turned on. It contains toxic heavy metals. It has an explosive glass tube, and we think you should put it at eye level with your children and encourage them to play with it. Do you want this in your house? This item is a color TV. The same is true for cars. Why would anybody want to own three thousand pounds of hazardous waste? I think we have a design problem."

According to McDonough and his compatriots, a better strategy is to have the manufacturer take back the durable product when the customer is finished. Companies should be well set up for products designed for disassembly, reuse, and recycling of the metals, plastics, and all the components. Rather than a "cradle to grave" cycle, it becomes what he has coined a "cradle to cradle" cycle. The cycle is good for business because it builds consumer loyalty to the manufacturer.

The third product typology is unmarketables, items such as nuclear waste, dioxins, or chromium-tanned leather, all of which are highly toxic and essentially undisposable. "We think that the measure we have to apply to our work is: What is the effect on mother's milk? Studies in Germany have shown that mother's milk in Munich can be significantly more toxic than would be allowed to be sold on the shelf. We've heard of people in Eastern Europe whose bodies are too toxic to be legally buried in a landfill. Dead whales found in the St. Lawrence seaway are too poisonous to bury in a landfill. This is how we're destroying biodiversity. We're dripping poison. We are making products or subcomponents of products that no one should buy. My partner Michael Braungart proposes that these products should be stored in 'parking lots' with manufacturers paying a lease fee until people find a safe way to dispose of them, and they should cease to be designed, manufactured, and sold altogether."

For the urban headquarters of the Environmental Defense Fund in 1986, McDonough created a "miniature university, an Athens using Spartan means." Believing that people would rather work outdoors, he made the indoors as much like the outdoors as possible. He designed a boulevard lined with trees to inspire the fund-raising staff. Every staff member has a street lamp and a tree nearby. There is an area called the pine grove "where the philosophers can discourse." Using special energy-saving light fixtures, the ceiling is rendered as the sky. The construction specifications minimized toxic chemicals and provided 30 cubic feet of fresh air per minute per person when the national standard is five. He also used materials like the natural fiber jute to replace synthetic carpet-backing materials, and insisted that it be nailed rather than fastened with toxic glue to the floor, thereby eliminating toxic off-gassing from the glue and synthetic carpet fibers. Built for a very low $38 a square foot, the EDF office demonstrated that a "green" office doesn't have to cost more.

THE ECOLOGY OF STUFF

In his tenacious quest for safe and sustainable materials, McDonough nearly fell off his chair scouting for alternative furniture materials. "We were asked by the world's largest office furniture company to design furniture fabrics," he says, fingering a coarse, elegant piece of material, deep green streaked with orange and yellow like an autumn forest. The pattern is actually a fractal abstraction of white ash tree bark. "We found that not only do we have to design what it looks like, we have to design the thing itself. They said, 'Wouldn't it be a great ecological fabric using cotton and PET, recycled plastic from Coke bottles?' We said, 'No, cotton as typically grown is responsible for about nineteen percent of the world's pesticide use, is irrigated in ways that damage local ecosystems, and has never been associated with social fairness. PET is an industrial-derived petrochemical that is not solar-driven or designed to be attached to your skin. You get neither a consumable product nor a product of service. Once you put the two together, you'll never get them apart. This is not a product that should be made." McDonough characterizes PET being reused in a park bench not as recycling. but *downcycling*, creating further product generations still in a cradle-to-grave life cycle destined for the landfill.

McDonough, Braungart, and their manufacturing team devised an elegant alternative. "The fabric we designed uses ramie from the Philippines and wool from New Zealand as the base fibers, organically derived. Wool absorbs moisture, and ramie wicks moisture off, positive qualities recommended by people confined to wheelchairs. It's heavy-duty industrial designing right down to the molecules. Then we went to the chemical industry and asked for process chemicals that met our criteria for safety." The response was uniform: silence. The designers then asked for access to the industry's chemistry to ascertain the safety of the basic chemical components. All the chemistry departments of all the companies refused, until they reached Ciba-Geigy, a huge multinational chemical conglomerate headquartered in Switzerland. They implored, "This is the future of chemistry; let us in." The company acceded. McDonough and crew then began examining fabric-related chemicals.

McDonough made the instructions simple: "No mutagens, no carcinogens, no persistent toxins, no heavy metals, no bioaccumulatives, no endocrine disruptors. Period." They scrutinized over 8,000 chemicals. Some 38 passed through McDonough's design filter. "This is safe stuff," he says, licking the material. "You could eat it. It becomes compost. You're can buy this fabric, and when you finish with it, it comes off the chair and you throw it in the garden. A consumer product that is consumable: What a concept!"

McDonough was delighted to receive a call from BSW Architects to collaborate on a new Wal-Mart in Tulsa, Oklahoma. The mammoth retail chain, which for a time was building a new store every two days, wanted to begin to explore environmental options. Since commercial buildings such as Wal-Mart have a short life cycle, 40 years or less, what becomes of the structure once its present use has run its course? McDonough advised them to design their one-story building with elevated ceilings to permit its conversion to two floors of apartments when the store completes it incarnation. He made sure there were no chlorofluorocarbons, the conventional cooling agents that cause ozone depletion related to climate change.

McDonough also convinced the influential retailer to use sustainably harvested wood in its roof for reasons including energy efficiency. To use steel in buildings can require 300,000 BTUs of energy per square foot, and concrete can take 200,000 BTUs, whereas wood can use only 40,000 BTUs. Also concerned about deforestation related to wood usage, he investigated wood sourcing, actually starting a nonprofit advisory organization to research sustainable forestry and woods. Woods of the World, a forest research information system, has now investigated the range of lesser utilized tree species and today boasts a database of three thousand tree species, with thousands more due on line soon. Some of the wood for the Wal-Mart store came from a 140-acre forest in Oregon, and its harvest was monitored by a local environmental forestry group to assure sustainable practices. The further use of engineered wood in the roof instead of normal timber beams reduced 30 percent of the wood needed, proving that there is indeed room at the top for sustainability. The Woods of the World database is now available to architects and builders.

Working with Andersen Corporation, the manufacturing giant, McDonough conceived the incorporation of six experimental skylights, which are the first to bring daylight into the otherwise windowless Wal-Mart design format. The special skylights, coated with holographic films that spread light, reduce the need for electric lighting and require fewer units than would otherwise be needed. The technology helped Wal-Mart reduce its utility needs by a bright 54 percent and substantially increased sales. McDonough estimates that this type of client alone could demand half a mile of skylights a day for five years, potentially birthing a new solar illuminating industry.

UNEMPLOY TECHNOLOGY; EMPLOY PEOPLE

Never one to think small, McDonough is already applying this inexorable logic of sustainability to society at large. His scenario of miles of skylights begs the question of a "closed-loop" industrial system that generates massive employment. The availability of new high-tech solar superglazing for windows provides an optimistic program for redirecting industrial economics. "We have glazing in windows that now can produce a positive net heat gain in winter even on the north side under certain conditions. It's called superglazing. Technologically, we're incredibly sophisticated. Have we figured out how to use them yet? No. If all you have is a hammer, everything starts to look like a nail. Our research staff is developing some very complex modeling using computers. How do we make our cities efficient? What about *superglazing the city?*"

According to McDonough, urban homeowners are not conserving energy because they are not encouraged to do so. People are loathe to spend up-front cash on special windows that would take 25 years to pay for themselves, even though the long-term energy savings are substantial. McDonough proposes public-private partnerships between the city and a window manufacturer where the city gives homeowners a financial incentive such as a tax rebate to have the windows installed. The company manufactures the windows employing local people and delivers them at one-third the conventional cost. The municipal program performs an energy audit house by house, does a cost/benefit analysis, then installs energy-efficient improvements including superglazed windows, and takes back the old windows for recycling. The payment on the windows is keyed to the homeowner's taxes over a 30-year period, and the annual payment on the windows is guaranteed to be less than the energy savings. Homeowners get energy-efficient housing, the city gains a window factory employing hundreds of people in new jobs, and the municipality leaves its children with an efficient metropolis. Citizens would then stop spending $250 million a year for petrochemical fuel, and the money stays in town with multiplier effects for the local economy. Economic modeling has to come about at the level of community, McDonough emphasizes, to build sustainable economic growth. The same scenario can be played out in cities across the country. "Remember," he states emphatically, "all sustainability is local."

"It's a highly complex model that can be a new way of thinking because we recognize our *interdependencies*," says McDonough with unrestrained glee. "Everybody has to be a designer. What we're calling for is massive creativity. The fundamental issue we're trying to address is the rightful place of humans in the natural world. How can we go about being part of natural design? The word *ecology* comes from the Greek roots *oikos* and *logos*: 'household' and 'logical discourse.' *Economics* comes from the same root. It is imperative to discourse about our Earth household."

Sustainability must inherently include passing the knowledge on to the next generation. So when McDonough won an international competition for the design of a day-care center in Frankfurt, Germany, he was intent that it have as much daylight as possible, allowing the children to become "attuned to the natural rhythms of the day." He further designed the roof with track-mounted shutters, which the kids can open and close depending on the weather or their needs. By being actively engaged, the children learn about natural cycles, and they now enjoy "putting the building to bed" each day. The solar design also has reduced the building's reliance on fossil fuels to a minimum. To gain auxiliary utility from the solar water collectors and cisterns, the building design stipulated a laundromat where parents can do their washing by the grace of the natural solar and water flows while waiting for their children to put the building to bed.

The overarching consideration in McDonough's work continues to be sustainability. Attending Yale for his M.A. in architecture in 1976 at a critical cusp in environmental history was an influence. "I got into architecture in the early seventies during the energy crisis. It disturbed me that the best designers in the profession weren't paying any attention to it. Those who were paying attention got ostracized by the high-design people as being junior woodchucks. I grew up in Hong Kong, and during the dry season we had four hours of water every fourth day. Resource conservation has to be seen as part of any design quality."

Yet few people have grappled with the specifics of what constitutes sustainability. Contacted to write the official design principles for EXPO 2000, the World's Fair for the year 2000 in Hannover, Germany, McDonough proposed that the event, rather than celebrate humanity's "conquest of nature" as it has in the past, instead serve as a working model for sustainable design. The city commissioned McDonough to create "The Hannover Principles: Design for Sustainability." The principles embody environmental design codes for the planning and construction of the World's Fair. They include such tenets as "Insist on rights of humanity and nature to coexist in a healthy, supportive, and sustainable condition."

McDonough envisions the Hannover World's Fair as a global work-in-progress modeling state-of-the-art, nature-based, humanistic design. "In the big picture, what we see is that this supposed dialectic we've been dealing with between socialism and capitalism is a very dangerous thing. The problem is that there's a third leg. In design, you need a tripod to be stable. That third leg is another 'ism,' ecologism. Without that third leg, without understanding that our home is the basis for all these activities, we will again collapse."

THE DECLARATION OF INTERDEPENDENCE

McDonough indeed collapsed his New York office in 1995 to reconstitute it in Charlottesville, Virginia where he was awarded the prestigious post of dean of the School of Architecture at the historic university founded by native son Thomas Jefferson. Standing at the center of the elegant "Lawn," McDonough points with awe to the famed rotunda Jefferson designed. The domed building arcs gracefully at the head of the oblong commons, whose buildings ramble along the sides of the gently terraced green grass field. Old trees populate the Lawn, sentinels to the mix of graduate students, faculty, and deans who scurry from the adjoining two-story buildings across the campus.

The trees, some of which were likely planted around the university's founding in 1820, are stately and diverse. Oddly, yellow ribbons are tied around many of them. A closer examination reveals Dean McDonough's penetrating influence on the students taking his class in record numbers on environmental choices and sustainability. Inspired by the dean's redesign of society, the students' yellow ribbons do not signify ritual compassion for prisoners of war. One large tree is marked beneath the ribbon with a sheet of statistics. It is a white ash over 150 years old and 100 feet tall. It is a grandfather tree that Jefferson himself may have passed by. The sign says it will yield 372,000 sheets of paper, or burn a light bulb for 10.9 hours. Perhaps it is a prisoner of war after all—the war against the environment. McDonough's students are seeking amnesty for trees from waste. They have asked the university to use only recycled paper and save energy.

McDonough's eyes twinkle, proud of this subversive information campaign the students have undertaken. He invokes Jefferson constantly as statesman, farmer, architect, plant breeder, philosopher, and framer of the Constitution and the Declaration of Independence. He ought to get a bumper sticker saying "Jefferson made me do it." He speculates on his plans there. "We're going to make the new Architecture School addition a net energy exporter. Students are going to find that if you want to draw on your computer, you may have to stop now because you left your monitor on all night, and although the sun was shining yesterday, you didn't pay attention to your power quota and you're out of juice. In the rotunda later this year, we're going to declare *interdependence*. Jefferson would do it if he were alive today. It's a declaration for life, liberty, and the pursuit of happiness free from remote tyranny. Except this time it's intergenerational remote tyranny, and the tyrants are us and our bad design. I think it's time for something dramatic."

In the aura of Thomas Jefferson, the University of Virginia may well be the breeding ground for yet another revolution, the design of the Next Industrial Revolution. The Iroquois Nation always planned according to the consequences for the seventh generation to come, McDonough observes. And his is the seventh generation. If he has his way, the Next Industrial Revolution will honor the seven generations to come.

CHAPTER EIGHT

Restoring Spirit

THE WORLD'S PEOPLE HAVE CONSUMED AS MANY GOODS AND SERVICES SINCE 1950 as all previous generations put together. People go to the mall more frequently than to church. Over 90 percent of teenage girls consider shopping their favorite pastime. In latter twentieth-century society, it appears that God is not dead. Rather, the market is God.

Perhaps the overarching phenomenon characterizing human life on Earth today is the sheer materialism consuming the planet. This voracious pursuit of material goods, branded deep into the scarred Earth, belies most traditional notions of spiritual pursuit. We are so weighted down with worldly baggage that it is hard to see how we could squeeze through those fabled pearly gates.

Humanity also stands at the Faustian threshold of what may well represent the deepest challenge to our values that we have ever faced: genetic engineering. This godlike manipulation of the genetic core of life itself calls into question our very beliefs about the nature of life and our place within it. How do we value life? Do other life forms have a right of life? Who's in charge?

Yet around the world there is deep concern about preserving basic human values and respecting the Earth. Many people are yearning for a restoration of some sort of spiritual relationship with the Earth and other living beings. Groups of indigenous people are raising their voices in an impassioned plea for peaceful coexistence with nature and reverence for life's diversity.

As many of our bioneers describe, all the technical solutions for environmental problems are insufficient to correct our straying course without a deep renewal of life-affirming values and an honoring of spirit.

207

THE PROBLEM

Our relentless assault on the natural world may be fueled by as trivial a pursuit as shopping. Picture this typically mundane scene played out daily by tens of millions of people: an air-conditioned car cruising through a drive-in fast food outlet for a burger and fries in disposable packaging. This seemingly innocuous activity exemplifies the key components that are among the most environmentally destructive in our lifestyle. The energy, chemicals, metals, and paper expended in this act represent four of the five most energy-intensive and polluting industries.[1] In a mere 40 years, cars have burgeoned from 50 million to 500 million. In 20 years, the number is projected to double to 1 billion.[2] Swing low, sweet chariot?

For teenagers, time at the mall is on a par with hours spent at home and in school. Just to supply newsprint to advertise those sales at the mall in U.S. dailies, Canada cuts primeval forests every year on an area the size of the District of Columbia.[3]

Does any of this afford us more leisure time? Quite the opposite. Modern workers labor much longer hours now than people did before the Industrial Revolution, and vastly more than they did in hunting and gathering societies, which spent a mere three weeks a year taking care of basic food needs.

The scale of consumption we are induced to buy into is mainly motivated by corporate greed, not by human needs. Are we possessed by our possessions?

Meanwhile, the gap between rich and poor is arguably the greatest in recent history. While most world spiritual traditions teach compassion for the poor, the afflicted, and the unfortunate, the meek are generally inheriting generational poverty and toxic waste zones. Is the fetishism of the commodity shredding the very fabric of human relations?

According to many traditions, the Earth was entrusted to humankind in a spirit of stewardship. Yet increasingly we are "playing God" by making fatal choices for countless other species and life forms. Evident in the precipitous loss of biological diversity— extinctions are expected to take one-fifth of Earth's species in just 20 years—our behavior raises spiritual questions about how humanity is serving.

In most spiritual traditions, all living expressions of the Creator have an intrinsic right of life. Yet geneticists are conducting advanced experiments in laboratories to violate species boundaries by moving genes around like checkers, inserting firefly genes into tobacco and flounder genes into soybeans. Human genes have now been engineered into the permanent genetic code of a variety of animals including altered mice, sheep, pigs, cows, and fish. At what point does an engineered life form cease to be sovereign and instead become an inanimate commodity? Where do we draw the line between life and things?

The Human Genome Project, designed to identify the totality of human gene functions by the year 2000, is leading to a race for patents on basic biological functions. United States corporations have already applied for patents on cell lines in women who are genetically engineered to produce lucrative biochemicals in their mammary glands. The engineering of other creatures including mammals has already produced monstrous outcomes, such as "superpigs" intended to be bred for lean meat, which instead turned out as "excessively hairy, lethargic, riddled with arthritis, apparently impotent, slightly cross-eyed, and barely able to stand up." According to author Andrew Kimbrell, "The new patent policy transformed the status of the biotic community from a common heritage of the Earth to the private preserve of researchers and industry."[4]

Agriculture is leading the way with indentured species tethered to corporate patents. Following close behind is medicine, as corporations are patenting human body cells, such as bone marrow stem cells which help create blood itself. Genetically engineered body parts are expected to follow, and even engineered human bodies. Some even contemplate the creation of personal clones, which could be used for organ transplants. Other scenarios project "people" reengineered to resist cellular damage from exposure to toxic wastes or other industrial hazards. Possibly the only legal restraint to such practices currently on the books is the Thirteenth Amendment on antislavery.

A disturbing survey revealed that 11 percent of people said they would abort a fetus that held a genetic trait for obesity. Who will make these choices? Who will decide what is a "bad" gene? Nazi doctors also attempted eugenic manipulations and showed the horrors that can result.

The patenting of life forms represents the ultimate commodification of life, calling into question the sacred nature of life itself. As human beings, we must reconcile our spiritual belief in the sanctity of life with our expanding facility to manipulate the very core processes of biology. But can we so handily tease matter from spirit?

SOLUTIONS

There is no word in Native American languages for the love of an object. Native cultures consider infatuation with an object a form of madness. Shifting our priorities from commodities to life may be the central spiritual challenge of our time. Spiritual leaders of all quarters agree that rampant materialism is impoverishing our spirit, and they find large congregations seeking healing and deeper meaning to their lives.

As is clear from the work of the bioneers described in this book, redesigning our production systems can provide abundance and actually improve our quality of life without plundering the planet. The widening division between the haves and have-nots is not intrinsic to our resource base. But do we really want to encourage our current levels of material consumption?

There is a global spiritual renaissance at the heart of which is the simple principle of reverence for life in all its forms. It is often founded on a new covenant with nature and a reaffirmation of compassion for all the world's peoples. Spiritual leaders ranging from the Dalai Lama to Iroquois Chief Oren Lyons to theologians Matthew Fox and Thomas Berry have raised their ardent words in defense of the voiceless creatures of the Earth. A lively debate is challenging the orthodoxy that humans are unique among earthly life forms in having souls. Eco-feminists such as Starhawk and many others from a wide sprectrum of faiths are calling for us to reintegrate ourselves as benign stewards within the web of life while honoring the sacredness of the Earth. The teachings of many indigenous peoples inspire us to remember that the Earth is alive and that all life is sacred.

The National Religious Partnership for the Environment is expressing the collective passion gathered from Judeo-Christian communities across the United States. Led by Dean James Parks Morton, the recently retired leader of the Cathedral of St. John the Divine Church in New York City, the group proclaims: "Care for creation is a fundamental religious duty, a sacred obligation to what God made and beheld as good." The National Religious Partnership, whose member groups serve 100 million U.S. citizens, calls for protection of the commons and preservation of the environment as "faithful stewards" for future generations. It also identifies this effort as one of social justice for the environmental harm especially affecting "poor and disenfranchised people."[5] Large church-controlled investment funds are increasingly screening their portfolios to avoid environmentally and socially destructive businesses, and to support ecologically oriented enterprises.

Pioneering medical researchers and authors such as Dr. Larry Dossey are documenting the efficacy of prayer as a healing modality, and finding mass acceptance for their work. This research accords with the interpretation of quantum physics briefly described in this book by Vandana Shiva, which implies nonlocal interconnections among distant phenomena.

Wherever women are empowered in communities around the world, environmental improvement generally follows. Women's environmental projects are emerging as among the most effective because in many cultures women have maintained the strongest respect for nature and community relationships. Restoring the feminine balance in spirituality is emerging as a vital force for restoring the Earth.

WHAT YOU CAN DO

Diminish your consumption. Examine what you really need versus what you may think you want. Think about the impacts of your consumption: Is it worth it? And is consumption (once the name of a disease like gout) filling up your life and crowding out other experiences? Set your intention and allow space for nonmaterial pursuits in your life, whether it's a walk in the woods, meaningful time with friends, or prayer and reflection. Look at the issue of service in your life. What are you doing for others, whether homeless people or your next-door neighbors? How are you supporting the natural world, whether birds or rocks? Are there things you want to learn? Are there people who could be teachers to you? Above all, make space for these changes to happen in your life, and call them in. You might be amazed at how quickly the richness of life may unfold about you.

As Prince Kropotkin maintained, and as some evolutionary biologists today agree, nature is indeed crafted to a large degree from mutual aid and often guided by cooperation. While fierce competition certainly exists in the natural world, there are also many instances of altruism. Yet it is not a value that we encourage or teach. Our society encourages a very self-centered life, and many people find that getting outside yourself is a richly rewarding experience.

As these bioneers describe, there are many paths to finding the integration between our inner life and the realms of outer service. Jennifer Greene and John Perkins respectively have journeyed deep into the spirit of water and the heart of shamanism to bring back visions of the relationships among culture, nature, and spirit. They suggest that it is time to invoke the grace of nature to redeem our souls.

Jennifer Greene's Fish Heart

Jennifer Greene has immersed herself in the spirituality of water. She finds profound spiritual lessons in this archetypal medium, nature's great channel of oneness and the collective unconscious. In this sacred element of nature, she reveals a magical dance of cosmic forces played out through this remarkable vessel of universal dissolution that is so forgiving, yet powerful enough to wear down a rock over time with its watery persistence.

JENNIFER GREENE IS IMMERSED. THE GENTLE, PORTLY WOMAN, PERCHED ON a stool in a wooden shed adjoining her converted barn-office, methodically drips drops of blue-tinted water into a large beaker of water on a workbench. Here in Maine's evergreen north woods, themselves a vast wetlands, she is rapt, observing the Neptunian patterns the blue drops make as they merge with the fluid world. The thick lenses of her glasses look liquid, dissolving the boundary between her and her watery fascination. Each drop cascades into a sequence of swirling forms. A ring vortex forms, only to transmute into a rotating funnel of spiral sheaths as she stirs it ever so slightly. It distends into a fanciful image of a sea creature in a fluent dance, until finally the blue permeates all, dissolving again to formlessness. Like a fish, Greene is one with her element.

"Water is not just two gases, hydrogen and oxygen," Greene says with gentle contemplation, while the sound of water running through an odd sculptural fountain pervades the space with dynamic calm. "Water is sentient. It is open—to good and bad. It accepts everything—hence our pollution problems. It gives up form, and takes the form of whatever you put in it. It becomes a vehicle for the formative processes of organisms. We are born into water. With water we are always moving back and forth between the worlds of the spirit and the senses. If we becomes students of water, of the wateriness of water, the very nature of water, we will learn a new social ethic. My quest is to discover the nature of water."

"FORM IS MOVEMENT COME TO REST"

Greene notes that the nature of water is movement, and quotes the Greek philosopher Heraclitus, who said, "Form is movement come to rest." She describes what happens as a blue drop starts to form on the tip of her dropper and move on its journey into the beaker. "When water drops, it goes through amazing metamorphoses. Every drop that falls first hangs, held by itself by this magnificent thing called 'surface tension.' Nobody really knows what surface tension is, but the outermost layer of water itself is different. When a drop increases in size, it fills up like a balloon filling from within. It begins to hang more deeply and more deeply and more deeply, until there is hardly a hair thickness of water holding that drop. Then it lets go, and flattens out on the top ever so briefly. We've been able to photograph a whole series of capillary waves over the surface of that drop. Each one of those layers is moving over the layer beneath it and above it with multiple motions between those laminations of water. If you were to peel a drop like an onion and lay it on the ground, it would cover half an acre."

The blue drop plummets through space, obedient to gravity. "The drop becomes an egg-shaped vertical form at one point," she continues, "and then becomes an egg-shaped horizontal form, and back again. When the drop falls into a body of water," she says wondrously watching the blue drop explode softly into the tinted liquid, "you have this crater and then this ring above it. A jet begins to emerge from the center of the crater. Then a sphere forms at the top of the jet and separates to become an independent sphere. The only place that a sphere forms absolutely perfectly is in outer space. The sphere is the archetypal form water wants to take. Why? Because it is the most holistic continuum of movement in water. When you look inside this jet, the impression is that it is being lifted. Water is the archetypal mediator. It mediates between gravity and levity, between stillness and motion, between life and death."

FLOW FORMS OF NATURE

Indeed, as Greene intently observes the endlessly changing form of the blue water, she sees a repetition of patterns, shapes that are mirrored in

214

the waterfall fountain whose hypnotic murmuring gives a soothing white noise below her soft voice. The fountain is a descending series of bilaterally symmetrical vessels whose shapes echo anatomical forms — vertebrae, heart, lungs. The water streams down in repetitive patterns, creating twin vortical eddies on a downward voyage like a mountain stream.

The fountain is actually a "flow form," a specially designed sculpture modeled on the natural patterns that water takes in the wild. In fact, Greene says, these natural forms serve to enliven and revivify water. The movement that is part of water's nature embodies distinct rhythms and pulses that support the purification process of water.

Greene became acquainted with flow-form technology in the early 1970s through her interest in Rudolf Steiner, the Austrian metaphysical philosopher, and the education she had received as a child at a Steiner Waldorf school. She read the extraordinary book by Theodor Schwenk, *Sensitive Chaos: The Creation of Flowing Forms in Water and Air*, many of whose ideas were based on Steiner's "Spiritual Science," known as anthroposophy, which views the Earth as permeated with divine and cosmic spirit. She would later meet the heirs of the flow-form work and bring it to the United States.

Flow forms were first developed by John Wilkes, a British sculptor and mathematician who worked with Schwenk in Germany. Schwenk was a hydro-engineer and student of Steiner who witnessed the declining state of the world's waters and vowed to restore them to a more natural state. Throughout Europe and the world, he saw how people had manipulated water, making channels and straightening rivers that had meandered freely since the beginning of time. He wept at the ever greater quantities of human and industrial wastes clogging the once sacred element.

"In olden days, religious homage was done to water," Schwenk wrote, "for men felt it to be filled with divine beings whom they could only approach with divine reverence. Divinities of water — water gods — often appear at the beginning of a mythology. Men gradually lost the knowledge and experience of the spiritual nature of water, until at last they came to treat it merely as a substance and a means of transmitting energy."[6]

Schwenk's intensive study resulted in *Sensitive Chaos*, a book Jacques Cousteau admiringly called the "first phenomenology of water." Schwenk codified the remarkable properties of the fluid element. He found water to

215

be the universal element, "not yet solidified but remaining open to outside influences, the unformed, indeterminate element, ready to receive definite form." He observed certain "archetypal forms of movement" in all flowing media, including air. He documented the universal patterns he observed, the funnel and vortex shapes that appeared not only in moving water, but also in clouds and in the morphology of plants.

LIQUID LAWS

Schwenk found several laws of flowing forms in water and air. As Greene concurs, water always tends to take on a spherical shape. It always seeks a lower level, drawn downward by gravity. Gravity pulls it toward a more linear path, but water invariably seeks to return to a spherical form. "A sphere is a totality, a whole," Schwenk observed, "and water will always attempt to form an organic whole by joining what is divided and uniting it in circulation. It is not possible to speak of the beginning or end of a circulatory system; everything is inwardly connected and reciprocally related. Water is essentially the element of circulatory systems."[7]

Greene points to the remarkable quality of levity that water bears. A phenomenon called the Principle of Archimedes allows a very heavy stone lifted from a river to be lighter while in water. Water's capacity to make bodies lighter is what allows it to rise vertically 80 yards or more through a pine tree, overcoming gravity.

Water moves through solid, liquid, and gaseous phases. Similarly, water uses hot and cold for its movement. Frozen water is hard as a rock, yet paradoxically ice floats on water by becoming lighter. If ice were heavier than water and sank, all the world's waters would fill with ice from the bottom and the planet would be a lifeless frozen ball.

Water is everywhere in movement. The ever-flowing oceans make up 71 percent of the Earth's surface. Plants also play a crucial role in water circulation. An acre of woodland on a summer day passes a 3,500 gallon stream of water into the atmosphere. Plants are vascular systems that pump water, "the blood of the Earth" as Schwenk termed it, in dynamic interaction with the atmosphere.

Schwenk found that water never flows straight ahead, but always travels in a winding path and from side to side. He discovered a constant

216

rhythm of meanders threaded with a dance of finer motions and inner currents. The twin movements of revolving circulation and downstream motion bring a spiraling action. These strands of movement are whole surfaces that interweave spatially and flow past one another. This rhythmical meandering is the natural path of water. "Everywhere liquids move in rhythms," he wrote. "Countless rhythms permeate the processes of nature."[8]

Schwenk wanted to know if the natural characteristics of water could teach something about how to repair the dead and dying water of European cities and landscapes. He founded the Institut fur Stromungswissenschaften (Institute for Flow Sciences) in the Black Forest area of Germany in 1961, and set about observing water with mathematician George Adams, Wilkes's teacher. He asked Wilkes to make models of these forms of "path curve surfaces" in hopes of revivifying the Earth's waters. The work is now carried on by Schwenk's son.

VORTICES, PULSES, RHYTHM, AND REJUVENATION

When Wilkes began to build replicas of naturally occurring water forms, he observed a pulsating meander caused by the water hesitating as it passed through a reduced aperture in the form. The hesitation created alternating channels of water, forming twin vortices rotating in opposite directions in a figure-eight form. Wilkes called it a "vertical meander," noting an unmistakable pulsing action.

When wild water moves from the headlands, Greene confirms, it is gathered from the peripheries through tributaries, rivulets, and springs. At this phase, it is filled with air and light and is usually fast moving. As it reaches the lowlands, meanders develop and deepen, and the water is slower moving, less filled with air and light. Its movement creates its own brakes, periodically digging up little troughs to slow itself, like the washboard of sand on a beach.

Schwenk conducted experiments on polluted water in the Black Forest to test his theory. "Schwenk found a direct correlation between water movement and water quality," Greene says, pleased with the discovery. "Water rejuvenates itself through its own natural movements. Within living creatures, life is maintained primarily through the pulsing of blood. Preliminary observations show water treated rhythmically by the flow form generates healthier and larger plants. Fish and other animals, including microorganisms, seem to thrive with flow form–treated water. Some flow forms are efficient oxygenators, and therefore enliven water. Schwenk understood that water is really a food and must have vitality." Water deprived of its natural movement is devitalized, Greene concludes.

The first experiment treating water with a flow form occurred in Jarna, Sweden in 1973 at Semeneriat, a Rudolf Steiner school. School director Arne Klingborg wanted to address the school's need for sewage treatment while providing an aesthetic and educational environment. The system encompasses seven ponds about 60 by 80 feet and four cascades of flow forms to service the 200 faculty and students and 15,000 annual visitors. The ponds successively evolve into more advanced ecosystems. The first pond harbors the most primitive microbial life and then progresses into ponds with algae, water fleas, fish, and other animal life. Finally the ponds become a water garden thriving with lilies, irises, and ducks. When the water flows out to the Bay of Jarna and the Baltic Sea, it is of a very high quality.

Simple pumps recirculate the dirty water through the sequence of flow forms. Whereas the water would normally take three months to digest the organic matter in the first pond, the flow form accelerates the purifying period to a matter of weeks. It appears to do so by oxygenating the water and thus supporting and quickening the microbial digestive process. Greene believes also that the rhythmical pulsing itself acts to cleanse the water. She points out that such rhythms are observable under a microscope in the microorganisms, which are continuously pulsing and spiraling.

The Swedish facility uses several flow forms, which mimic different stages in the process of water flow. The "Jarna" form is a repetitive figure-eight, which resembles a series of vertebrae, and provides the vigorous movement that Greene says is especially useful for wastewater treatment. Other forms provide widening and narrowing shapes that hold the water longer and produce greater rates of hesitation, pushing the water into vortices and higher pulsation rates. The forms often match shapes similar to human internal organs such as the heart and lung, the organs that regulate circulation and move the blood.

Greene's fascination with water began early. "I played with mud puddles as a child in Vermont," she recalls with a sly smile. "The greatest laboratory for flow phenomena is the mud puddle in spring. The silt that has been dissolved through the freeze-thawing process on the ground seems to be a perfect indicator of subtle movement patterns. You have it all right there—back eddies, trains of vortices, streams moving within streams, upwellings and ripples and riptides. When spring comes around and there's a tiny little trickle into a puddle, I'm out there with a video camera."

Growing up in the Vermont countryside, Greene attended a Rudolf Steiner Waldorf school where a botany teacher enchanted her with the beauty of growth and form. The education devised by Rudolf Steiner also emphasized eurhythmy, the study of rhythm and movement through language and tone. Her interests continued in college at Sarah Lawrence with biology, physics, and sculpture, and as a freshman she came across Schwenk's *Sensitive Chaos*, which so captured her imagination.

In 1973 she made a pilgrimage to Dornach, Switzerland, a Steiner center, where she met Schwenk's colleagues and learned about the other work the Institut was doing on studying water quality. In 1980 she built her own lab to carry on the work in association with the German Institut.

DROP-PICTURE MANDALAS

Schwenk had developed a photographic process for examining water called the "drop-picture method," which Greene started studying. Inside her office, she proudly displays the odd-looking contraption, a double optical bench. A camera is positioned on a tripod at one end of the upper bench, while at the other end water drips slowly into a shallow dish containing the sample. The light source at the bottom and a series of calibrated lenses afford a glimpse into what goes on between layers of water. The camera takes periodic snapshots of a sequence of thirty drops as they fall to the surface below, revealing extraordinary mandala patterns in the pictures.

"Schwenk really has opened our eyes as to how sensitive water is," Greene says thoughtfully. "He had to find a way that would describe differentiations in water quality when the samples of water, which were the same in terms of chemical and biological activity, were moved in different rhythms and over different models. The drop-picture method gives a pictorial image of water quality."

The teacherly Greene presents two drop-picture photographs from her extensive collection. The differences are easily visible. One picture shows a radiant mandala with tendrils extending like an exploding star or a sand dollar. It has the pattern of a rosette. The other picture is contracted and depressed, a shadow of the former. It lacks the rosette imprint.

"The latter picture," Greene explains, "is polluted water. What we have found is that the drop-picture method shows any disturbances in water quality sooner than most instruments." She points out that the common method of using a tensiometer, which measures the surface tension of water, to detect pollution is far less sensitive. A tensiometer will begin to register readings around one in 100,000 parts, whereas the drop-picture technology can see pollution at one in 500,000 parts. By the time the pollutant reaches high concentrations of one part per 50,000, Greene finds the surface tension so disrupted that no patterning shows on the boundary surfaces, indicating lost vitality.

She suggests trying a home experiment of first dripping drops into a pan of water and watching the capillary waves, followed by adding a tiny bit of detergent in the corner of another tray of water for comparison. The detergent stills the water flow dramatically, and, according to other experiments Greene and her European colleagues have performed, the common detergents used in copious excess by society are all but killing water. Use of the drop-picture photography in Europe has led to much stricter regulation of detergents. When the reservoir water supply of Stockholm, Sweden, a city surrounded by pristine conifer forests and reputed for its good water, inexplicably showed some of the worst water quality in Europe, the problem was traced to a leaking pipe in a laundromat. The result was a revolution in the European soap industry.

Greene believes that the drop-picture method can illuminate a different way of diagnosing water quality altogether. "By definition, we know water as good when it is colorless, tasteless, and has nothing objectionable in it. But we don't know what good water is. The drop-picture method gives us another view." Greene has recently been working with Trinity Springs, a natural underground spring in Idaho that has tested for the best water she has ever seen. She hopes to use it as a benchmark for establishing positive criteria for water quality.

Greene has been patiently waiting for the precise moment of four in the afternoon to take another drop picture. The reason for the exact timing, she explains, is that water is highly influenced by planetary forces, which can skew the pictures, just as the moon turns the tides. She and her colleagues have found that certain "interpenetrating resonances" occur when planets move in specific angulations. At those moments, she says, "Water is our true astronomer. It is as if things become open and water will pick them up. We ought to be able to see the workings of planetary movements in the water. Water brings what is of spiritual and cosmic nature down into the material world."

THUMBPRINT OF THE COSMOS

Greene learned first from Schwenk's work that impressionable water bears the thumbprint of the cosmos, of "higher forces penetrating through it into the material world and using it to form the living organisms." Schwenk went on to posit: "The world of the stars permeates all movements of water; that water infuses all earthly life with the events of the cosmos. Thus water becomes an image of the stream of time itself, permeated with the rhythms of the starry world. Cosmic events and water are linked with the stream of time in the ebb and flow of tides."[9]

Greene's nonprofit Water Research Institute now works closely with Schwenk's Institut on drop-picture photography, the only such collaboration in the United States. After her initial visit to Europe in 1973, she invited John Wilkes in 1980 to the United States to discuss a collaboration. They decided to form a company, Waterforms and Associates, to design, manufacture, and market flow forms. Working with molds, they have created a series of revivifying sculptures for use in homes, gardens, parks, schools, and institutions. She is also developing a school curriculum on water.

In Keene, New Hampshire, Greene's company put a flow form into the garden atrium of the Cedarcrest Handicapped Children's Home, and found additional benefits to the watery sculpture. "These children are tiny little

221

fellows, and when they're fussy and cry, the staff brings them to this flow form and they quiet right down."

Greene finds that water is a kind and valuable teacher for children altogether. "Kids turn off to environmental issues unless there's a positive message about what you can do. Children know what fun it is to play with water. So we are developing water play parks for urban areas with colleagues in Germany. They have all kinds of sluices and gates, a water castle and vortices where you can see a myriad of phenomena. The youngster below the sluice has to do some work so that the youngster above can get some water to play with. They have to talk to one another, to communicate. They have to judge how much water is needed. It requires the use of social skills and a lack of egotism. This park can be filled with a hundred children, talking with one another, managing the water, sending little boats down the river. It is absolutely vital for the education of our young."

Greene placed a flow form in a housing development in whose courtyard it "muffles noises, brings rhythm back into the auditory experience and becomes a gathering area for community. Wherever flow forms are, people gather. The flow forms are a kind of open heart." Greene has also installed flow forms in a park in Santa Cruz, California, which helps remediate the water in a pond used by hundreds of ducks that nest and play there. She is proud of a park in Germany that has a meandering river running through the old cobblestone street past a cafe and into a wetlands area. She is working on an environmental barge in the channel adjoining the Children's Museum in Boston. "We were asked to deliver water from the mezzanine down to the Wateriness of Water play park. At the bottom there's a form based on the fluid flow of a fish heart. It awakens the sense of wonder."

HOLY WATER

For Jennifer Greene, the ultimate lessons of water are spiritual. "Water has a selfless quality. Water is inclusive: It engages; it absorbs." Water wears down all solid things, Greene adds. In perpetual motion, it is constantly re-creating and re-forming the Earth, endlessly transforming rock into finely sifted flows of matter. Quoting Schwenk, she invokes the eternal truth of water. "'Like an echo of the ever changing events of the heavens, the fullness of form in the world comes forth from water. Water pictures — as though in a great parable — higher qualities of humanity's development. Qualities such as the overcoming of rigidity in thought, of prejudice, of intolerance; the ability to enter into all things and to understand them out of their own nature and to create out of polarities a higher unity.'"

"We are not usually apt to defile what we first revere," Greene concludes, scanning the molds of flow forms in her workshop. "When we allow water to become our teacher, it has a lot to show us about egotism, and how to overcome the intransigence of life, to move, to change, to be open, to give way — things that are very hard to do. When we learn from the nature of water, we will become far greater stewards of the watery household of Earth and our own soul natures. Then we can heal the Earth."

Walking outside, Greene points to a flow form she recently installed in her grassy green Maine yard. The trickling sound is enchanting as the water cycles endlessly through the gracious fountain. Two frogs have taken up residence in this tiny gurgling flow-form sanctuary. Life is thriving, seeking its level in the common pool of the wateriness of water. Over a sea of green pines near the ocean, sunlight pierces the rolling fog to cast a rainbow, the misty gate of colored water arcing between heaven and Earth. Jennifer Greene's fish heart is happy.

John Perkins:
Changing the American Dream

John Perkins works with shamans and native healers to effect Dream Change.
He leads tour groups into the Amazonian jungle to study with these indigenous
teachers and help conserve the rain forests. He founded groups in a loose
global network to spread the message of the shamans and promote
Earth-honoring wisdom.

AS THE HEALING CEREMONY FOR THE NORTH AMERICAN VISITORS CAME TO
a close, Manco, the Quechua Indian shaman, motioned for John Perkins
to stay as the others filed out, exhausted from the all-night ceremony.
Perkins was hesitant, concerned about the well-being of the group he had
brought from the United States to learn from the Andean and Amazonian teachers. But he knew he could not refuse the shaman's request.

Manco looked impossibly familiar to Perkins, although he felt certain
he had never before met the native elder. Despite conducting an extensive six-hour healing ceremony, the 103-year-old man remained spry and
energetic. The strange sounds of the wild jungle pierced the night as the
shaman took a candle and illuminated a small wooden trunk on the floor.
Handing the candle to Perkins, he opened the trunk and removed something. Perkins was stunned by the small leather pouch Manco thrust in
front of his eyes.

Perkins had seen the pouch before during intense recurring dreams he
had while attending Middlebury College some 25 years ago. They were
dreams of a young Incan running through high mountains, holding the
leather pouch and strings of knotted, colored yarn. He also remembered
distant dreams of corn, a gigantic bird, and an old shaman living high in
the Andes.

"Have you ever heard of the Birdpeople?" inquired Manco, a trickster
watching Perkins's unfolding revelation. Perkins had indeed seen a Birdpeople ceremony during his Peace Corps days among Quechua Indians, an

elaborate dancing circle of initiates wearing feathers and giant condor wings. "It is the job of the Birdpeople to remold the human community," Manco stated.

Perkins's dream, Manco explained patiently, was of the relay runners who unified the four corners of the large Incan nation by carrying messages made from knotted colored strings of yarn, secret codes intelligible only to the *quipu camayoc*, shamans trained to interpret them. They were taught to *camay* the yarn, or breathe wholeness into it. "I think our species needs some good *quipu camayocs* now," Manco confided, reading Perkins's startled eyes.

CHANGE THE DREAM

Holding the pouch, the shaman poured its contents into his hand, revealing golden kernels of a traditional Incan corn. Taking a mouthful of *trago*, a corn liquor, he blew a streaming mist over the candle, *camaying* a crackling spray of fire. He began chanting softly. "The seed is the dream of what is to be. Remolding requires only that we change our dream. For this we must plant new seeds."

Stunned, Perkins listened distractedly as Manco said, "Come to me to help your people. Come any time." Perkins, the shadows lifting from his memory, responded uncertainly, "I dreamed of you many years ago." Manco replied with assurance. "I had those same dreams. They came true. With both of us dreaming, how could they not?" Manco explained that his own teacher had given him these seeds, and now he was passing them on as Perkins's teacher.

John Perkins, a tall, earthy man with the green eyes of a jaguar, draws a long breath and speaks softly, choosing his words carefully. His story descends from the high Andean Quechua mountains to the Amazonian lowlands of the Shuar. "Like shamans throughout the world, the Shuar's belief is that the world is as we dream it, as we perceive and will it. Today the Shuar are familiar with what's going on in the rest of the world. A number of their children have attended high school in the cities and even universities. They're very aware of the environmental destruction, and they told me that this is what we have dreamed: the choked highways, air pollution, water pollution, the destroyed rivers and lakes. We wanted the material wealth, and it's all part of it.

"And they said, 'Only now are you beginning to realize that your dream is a nightmare. But it's easy to change that. All you have to do is change the dream, and everything else will change along with it.' I asked, 'How long will it take?' They said that it can easily be done in a generation. We need to change to an Earth-honoring dream."

RETURN TO THE FUTURE

Perkins's introduction to the Amazon occurred in 1968. Fresh out of college, he joined the Peace Corps. "I was sent to Southern California for eight weeks to train as a specialist in credit and savings coops," he recalls ruefully, "and then to Ecuador to this little village in the jungle to form one. It was absurd because once I got there the people looked at me with great shock since everything was done by barter. There was nothing to save! You can't horde bananas for very long. I didn't really have anything to do except learn about the jungle and the people. I spent a long time in the rain forest with the Shuar, getting to know them, and with the colonists, too, who were moving into the jungle at that time, taking land from the Shuar. I didn't do anything for the people there as a Peace Corps volunteer, but I learned a lot."

The nightmare of his own culture's impact came crashing down on Perkins during his return trip to the Amazon jungle 23 years later. "In the Shuar village surrounded by forests where I once lived as a Peace Corps worker, now the road is the center of activity. From the village, you can't see any jungle. It's been destroyed, turned into cattle ranches and other kinds of farms which aren't working. The people who went out and colonized are no better off than they were originally. They just have to keep moving farther out.

"But I had the good fortune to hitch a ride on a small single-engine plane to a village that's another five days by foot where the Shuar still live in virgin jungle. This village looks like the one I used to live in. It left a deep impression on me of the incredible importance of stopping that destruction of the jungle soon—immediately."

For the soft-spoken Perkins, the quest to save the rain forest holds a special resonance, because he believes he was partially responsible for its destruction. Following his years in the Peace Corps equipped with a degree from business school, he became a highly regarded economic consultant to institutions such as the World Bank, United Nations, Asian Development Bank, and Inter-American Development Bank, which initiated the ranching, oil, mining, and timber industries which have since devastated the ancient rain forests in the name of "progress."

Perkins eventually had nearly 50 people working for him in Asia, Africa, the Middle East, and Latin America. "As time went on, I became increasingly haunted by the problems that we were creating with these projects. After about ten years, I really began to wonder about what I was doing and what we were all doing. I had thought maybe I could change the system from within. I was particularly concerned over the total lack of regard for environmental and cultural issues. The whole driving force behind everything we were doing was to increase the gross national product of these countries with total disregard for how that was distributed, or what it did to the environment and society. Every time I brought this up, I was told, 'That doesn't have anything to do with what we're trying to accomplish here. What's important is the GNP per capita. That's the only measure we have.'

"Finally, I just got very frustrated and decided that working within the system wasn't the way for me personally. It was a really difficult decision because I was making a lot of money, having a very good time, traveling all over the world with lots of people working for me. It was a glamorous and prosperous job, and yet my conscience couldn't take it anymore. It still haunts me today, those terrible mistakes we made."

THE SHAMAN'S BRIDGE
TO THE OTHER WORLDS

Yet during his ten years living and working in so-called lesser developed countries, Perkins gained valuable insights into the indigenous cultures, spending much time with shamans and medicine people. "I'd become very interested in shamanism, primarily because this seemed to be a way that

227

people could help themselves with health and in other ways while also honoring the Earth and nature." Through them he learned shamanic practices and techniques of meditation that he calls *psychonavigation*, which he describes as the ability "to travel through the psyche to the place you need to be." Shamans use the method to balance the worlds of spirit, nature, and people. A shaman is considered a bridge, one who travels to other nonordinary worlds to use the powers from those worlds and this world to effect change.

Exposure to shamanic practices allowed Perkins to make more sense of some of his earlier life experiences. Growing up as an only child in New Hampshire, he spent long periods alone in the north woods, entranced by nature and often fantasizing about the Native Americans who once lived there. When he discovered from his grandfather that his family had indigenous Abnaki Indian ancestry, he became obsessed with Indian lore, especially the Iroquois notion that plants and animals were his brothers and sisters. He was moved by their belief that the preservation of the Earth is the highest calling. His parents discouraged his native self-identification, reminding him that "the frontier was gone," and that he was also related to Ethan Allen and Tom Paine, more acceptable heroes.

Nevertheless, in seventh grade Perkins wrote *Trail to the North*, a novel about the attempt of the Abnakis to prevent Europeanization. On the cover he drew an elaborate traditional bark longhouse in a forest clearing. When his teacher returned the paper with an "A," she also gave him a photograph of a longhouse almost identical to his drawing. It was not Abnaki, but Amazonian. "People still live like your Abnakis," she told the boy. Perkins smiles. "Dreams do come true. At that point I knew it was my destiny."

Following his departure from international consulting, Perkins undertook a large-scale environmental power project using waste heat from coal tailings. Several years later, he sold the model plant, highly successful both financially and environmentally, to Ashland Oil and again contemplated his life. "I wanted to get back to working with my friends in Latin America. I had been writing books and going down there, getting back in touch with these people. The whole carbon dioxide 'greenhouse effect' issue that I encountered in the power plant project was also very tied to the destruction of the rain forests. I wanted to go back to the people who had done so much to help me and now were having so many problems. At that time, about forty million acres of rain forest were being destroyed every year, an area the size of Florida and Connecticut combined."

Teaming up with his publisher Ehud Sperling of Inner Traditions International, Perkins headed south. The night before hopping the small plane that would take them deep into the jungle, Perkins tangled with some German travelers, arguing vehemently against dispatching blundering tourists into these fragile lands and peoples. Sperling admonished him that if he really felt that way, perhaps he shouldn't be going himself. It was a long night.

RIPPLES ACROSS THE MOTHER

Although he had walked through the rain forest many times years ago, this was the first time Perkins had flown over the jungle canopy. To the west he could see the devastated lands of the colonists, an ugly grid of incoherent clearings branded into the emerald skin of wilderness. To the east they flew over a seemingly endless mantle of forest green, which gave way to a clearing with an airstrip. There Perkins saw a large longhouse — almost exactly the image he had drawn in seventh grade. Despite the thick jungle heat, he shivered.

As the plane skidded onto the rudimentary runway, Perkins was surprised to see the contingent of Shuar men awaiting their guests. Unlike the headbands and loincloths of his memories, now they were wearing T-shirts and cotton shorts. He sat by the river examining their dugout canoes, bemoaning the materialism of his own homeland to Numi, a Shuar shaman. "You have lost touch with the Mother," Numi said. "Now it begins to hurt. You see that your dream is a nightmare. The problem is that your country is like this pebble," he continued, flipping a stone into the river. "Everything you do ripples across the Mother. If you want to help these forests, then you should start with your own people. It is your people more than mine who need change."

What Numi told him made sense. He asked the Shuar elder if he could help in any way. The answer came as a shock. "Bring people here," said Numi with a toothless grin. "We are masters of Dream Change. We can teach your people to dream change into the world."

Perkins was wobbly, recalling his argument with the Germans. "You want tourists in here?" Not tourists, Numi explained. "Bring people who want to learn. Mother Earth will survive. But if the people who call themselves 'civilized' continue to dream their greedy dreams, Mother Earth will shake us all off like fleas. Many other animals and plants will go with us. Many have already gone. The dream of your people must change."

Saving the rain forests was intimately connected with changing the consciousness of his own people, Perkins understood in that moment. Back in the United States, he had little idea what to do with his new instructions. More dreams came to him, especially one with an old shaman named Manco, who beckoned him. He decided to organize a group trip to Shuar country.

A series of synchronicities immediately began to cross Perkins's path. An old classmate from Middlebury College, a psychology professor who had read his books, called and then offered to organize a group to go into the jungle and learn with the Shuar. Perkins had joined the board of Katalysis, a nonprofit enterprise building North-South development partnerships, through which he learned about creating training centers for environmentally and economically sustainable projects in Latin America. Some centers were designed for people from the North to learn from those in the South about honoring Mother Earth. Within a year, Katalysis would receive a $2.5 million grant that included funding for such a center.

During the ensuing time, Perkins would lead three groups of a total of 35 people into the Andes and Amazon. These were followed by 12 college students, who each received nine college credits for studying with the Shuar and Quechua. "My objective was always the same," Perkins reveals. "The teaching was to have them experience their oneness—with nature, one another, and the universe." To reach Shuar country, the group boarded dugout canoes and braved shrieking white-water rapids. They trekked through deep forest alive with snakes and poisonous insects. They slept in rough lodgings in the heart of the jungle, where any emergency could mean death.

Yet they also experienced the grandeur of the land, Perkins says with exuberant laughter. "From the snow-capped Andes to the steaming rain forests, Ecuador is charged with sensuous energy. One hour you are awed by the incredible majesty of the world's highest active volcano; the next you are standing astride the equator watching a snow blizzard engulf gigantic Cayambe Mountain. People begin to accept the idea of an empathetic unity. The love stays with the travelers for a long time, perhaps forever." It was Perkins's hope that anyone who spent a day in this wonderland would not tolerate bulldozing the forest.

SONG TO A STONE

The groups got a firsthand experience of the indigenous worldview that sees the Earth as alive. "When a loved one is far from home, the two lovers send messages to each other through the Earth. A woman may sing a song to a stone. It is passed along to the next stone and on and on until it reaches her lover. He hears her song and sends his own message back in the same way."

On one of the trips, the appointed shaman was late, and Perkins's partner had an intuition that they should go to another shaman. It turned out to be Manco. During the session, Perkins witnessed remarkable healings. One was a medical doctor from Chicago with chronic neck pain. Another was a woman with painful herniated disks in her lower back. Both experienced instantaneous relief, not normally thought possible by Western medicine.

When Manco revealed himself after the session, Perkins knew his dream was coming true. Manco went on to describe a vision he had when the *gringos* first entered the room. What astonished the Quechua shaman was the sheer size of these people, many times larger than his own people. "May you dream of growing smaller," he chanted. "May your children require less clothing, smaller houses, less food. May they grow up to be my size and give more than they take from Pachamama." A tear formed in his eye for Pachamama, the Mother Earth and Universe.

As Manco explained the need to change the dream, Perkins asked how. Manco made a distinction between dreams and fantasies. He related his own fantasy as a young man of sleeping with his neighbor's wife, an obsession so strong that it occurred. The transgression caused terrible distress, however, and Manco wished the act had remained a fantasy. He said that while fantasies may be compelling, we don't really want them to come true.

Dreams, however, can change our lives when rightly applied. Manco related to Perkins that the people of the North must learn to stop turning fantasies into dreams, the first step in becoming a shaman.

"For us," Manco said, "prosperity is clean air and water, living close to Pachamama and our families, and eating fresh foods that we or our neighbors have grown with loving care. It is the knowledge, learned from daily living, that we are all one. It is honoring and protecting our Mother Earth, knowing that she is always here for us."

Manco went on to tell Perkins that the people of the Andes also once became possessed by greed. The male and female gods, Inti the sun and Mama Kilya the moon respectively, sent their sweat and teardrops to the people in the form of gold and silver. But the people became greedy and selfish. Then the Great Creator Viracocha got angry and sent his son Eagle to teach the people a new dream. Since that time, Manco's people have practiced the Eagle's teachings of dream change. To do so, they became Birdpeople.

Manco taught Perkins the means of dream change. The first step is to define what we want, to separate dreams from fantasies. The next is to give the dream energy, to give it voice and song. Finally, he explained, do not let anyone deflate the dream with negativity or doubt. No matter how impossible the dream may seem, Manco cautioned, giving it energy can make it real.

SHAMAN UNIVERSITY

For John Perkins, the task of changing the American dream is altogether possible. He has helped create the Dream Change Coalition, a grassroots network of people throughout the world dedicated to forest conservation and Earth-honoring consciousness. The group grew out of workshops Perkins conducts in the United States, Europe, and Latin America. He has recruited individuals, nonprofit groups, and companies.

Perkins linked the nonprofit company he founded in 1986, Prydwen (the Celtic ship of wisdom), with the Dream Change Coalition to expand its teaching activities, including trips to visit the shamans. He now works closely with entire professions, including medical doctors, psychologists, lawyers, and religious leaders. He is presently collaborating with the University of Houston School of Management to organize a program to take U.S. oil company executives to study with the Shuar and experience the damage oil exploration is causing in the rain forests. Since the oil companies are currently facing a $1 billion lawsuit from the Shuar, Quechua, and other Amazonian tribes, it would behoove all concerned to come to an understanding with the indigenous peoples.

Perkins and the Coalition have helped found two shamanic learning centers in the Amazon. One lies deep in Ecuador in impenetrable jungle with great rivers, freshwater dolphins, giant anacondas, and otters. A bird photographer recently filmed 87 species of birds in two days there. The second center is in a cloud forest in the Andes, 500 acres spanning five mountains that may be the most biodiverse area in the Andes, according to respected biologists. Because the owners of the land came under pressure to sell, members from one of Perkins's tours raised the money necessary to preserve it and create the center.

The Dream Change Coalition is organically growing into a loose global shamanic network. "It is a grassroots organization that just sprang from people who have been to my workshops and on my tours. We're trying to use the group to buy the land before it gets cut down. People are examining other 'productive' sustainable endeavors such as growing orchids and other ornamental plants within the forest as an alternative to oil or lumber. As a group, we get together to *camay*, to breathe life into a new dream."

"THE WHEAT MUST DREAM OF THE BREAD"

Perkins remembers returning from the natural bliss of the rain forest and landing bumpily in the synthetic shock of the Miami airport. During the weekend with his wife and daughter, they made a trip to the mall to replace his tattered jeans. Although people were materially prosperous, he was struck by the contrast of their frowning faces and sad eyes refusing to meet his smile.

"Our materialism is so seductive," Perkins says. "When you feel you want something badly, think of what it is that you really want. Is it really that new car or bigger house, or could it be the divineness within and the recognition of your oneness?"

But why, one wonders, has it been necessary to have this terrible nightmare of materialism? "If you believe, as the shamans teach, that the wheat must dream of the bread and the child must dream of the bread in order for the bread to be eaten by the child, then you must also assume that the Earth had this dream of being paved over. Why did the Earth have this dream of being so punished and so hurt? Perhaps it's like an infant moving from an abusive phase to one of more mature stewardship. I asked my eleven-year-old daughter the same question. Pointing to the stars, she said, 'Well, Dad, we're just a little tiny speck here in the whole oneness of it all. Maybe Pachamama had this dream so that the rest of the oneness could learn.'

"There's so much for us to learn about spirituality, about loving the Earth, about choosing ecstasy over materialism, about ourselves, about oneness. The indigenous people have a great deal to teach us. I see my role as helping my people, who are suffering such pain, to make these changes. We're like a small stone in a big lake. Drop us in, and despite our small size, the waves we create will cover the whole pond."

John Perkins is dreaming again of the rain forest, of rowing upriver. He knows life is but a dream, and we can change it.

Hearing the Vision

"CHANGING THE DREAM," AS JOHN PERKINS DESCRIBES IT, IS THE FIRST STEP in restoring the Earth. He echoes Black Elk's famous words: "Without a vision, the people will perish." The bioneers are creating such a vision of a renewed society in tune with the natural world, imbued with spirit and devoted to the reenchantment of the Earth.

At a Bioneers Conference in 1995, Paul Hawken recalled a scene from the movie *Close Encounters of the Third Kind* in which the Earth scientists first make contact with the extraterrestrials. As the scientists are playing tonal sounds to communicate with the several small craft visibly hovering around the skies, their excitement is mounting. But unseen behind them is rising the Mother Ship, inconceivably gigantic, an entire world unto itself. Hawken compared the Mother Ship to the supreme reality of the biological world. The wake-up call is inescapable. It is the overarching issue of our epoch.

The biological clock is ticking, and the more we resist and delay, the worse the pain and suffering will be, for us and for the myriad other travelers on the planet. The choice is ours. As we have seen, solutions clearly exist that are practical, cost-effective, and elegant. But what will it take to precipitate the changes?

The real and present threat of environmental catastrophe is certainly a compelling motivation. The sea change in the attitude of the insurance industry, making it a serious social force to avert the danger of global warming, is evidence of a societal shift. The dramatic changes in the medical industry, which is heeding both bottom-line incentives and the grassroots popularity of green medicine, is another dramatic example of deep social change. The high-level attempts among giant companies to redesign industrial protocols according to ecological principles are a further harbinger of the scale of changes that are rapidly approaching.

An exciting new field combining science with the art of nature is showing how we are beginning to learn to operate in harmony with natural design. Called *biomimetics*, the study of the structure and function of biological materials that can be used as models for human-created ones, it examines both underlying biological processes and their compositions. Extensive studies of such forms as reindeer antlers, sea shells, and spider webs have uncovered far more extraordinary properties of strength, resilience, and shock resistance than anything manufactured by people. Spider threads are stronger than steel and more durable than nylon, and these silks are teaching researchers about the "molecular architecture" embodied by the liquid silk secreted by spiders. "We have a lot to learn," commented one humbled scientist.[1]

But, perhaps the most poignant message of the bioneers is one of inspiration at the magnificence of the natural world and the extraordinary creativity of human beings. It is hard not to be awestruck by the beauty of the parallel universes these visionaries reveal. It was beauty that killed the beast, they say. Perhaps inspiration can motivate as well as fire and brimstone.

It is also essential to realize that the work ahead will occupy generations. It has taken a long time to bring matters to the dismal state they're in, and it will take time to set things right. Institutional change is almost always painfully slow. Changing material reality is a muddy road for thick wheels. It is not merely about changing our lifestyles, but changing our lives. The ultimate task is one of healing some very deep wounds and long-term chronic conditions.

For all the stories of bioneers in this book, there are dozens of others equally fascinating and inspiring. The Bioneers Conference has brought some of these people together, and the gathering has begun to spawn initiatives and collaborations that are leading to greater understanding and action. As awareness of these solutions spreads, more and more people are getting involved. Around the world, similar stories are taking shape. Social scientists have found that it takes only a small percentage of members of a society to make a revolution, as happened in the American Revolution. Restoring democracy from the current corporate stranglehold is no small part of the task ahead.

Taking personal responsibility through direct action is a step we can all take. We are all potential bioneers, capable of sharing the vision and acting. People around the world are making their voices heard in demanding a healthy environment. Successful models of authentic restorative development are gaining attention. The movement is expanding, and in many places people are seeking to take power back over their own lives and communities. You can vote with your life.

But viewing the terrible destruction and misery we have wrought, it is hard not to wonder what deeper darkness may dwell in human nature. Visionary biologist Lyall Watson, in an extraordinary book *Dark Nature: A Natural History of Evil*, explores the notion of evil from an ecological point of view and draws several germane conclusions. "If 'good' can be defined as that which encourages the integrity of the whole," writes Watson, "then 'evil' becomes anything which disturbs or disrupts such completeness. Anything unruly or over the top. Anything, in short, that is bad for the ecology." Good, he goes on to say, is "not necessarily the opposite of evil, but one part of the field in which both exist. That part which straddles a fine line along which things are 'just right.' "[2]

Watson defines three principles that together comprise the "Goldilocks effect." In the classic fairy tale, Goldilocks comes upon the house of the three bears, and tests each bowl of porridge and each bed until she finds one that is "just right." When the exquisitely "just right" balance of nature is disturbed, says Watson, scary things start to happen.

The first principle is that order is disturbed by loss of place. The right thing in the wrong place can be disastrous, like a plague of rabbits or kudzu vines in the "wrong" locale. Similarly, a people removed from their connection to a particular place can be catastrophic, as Jason Clay described.

The second pattern is that order is further disrupted by a loss of balance. Right and wrong numbers lead to bad things, whether too many or too few. Numbers and distribution are crucial in the ecological balance.

Finally, says the biologist, order is destroyed by lack of diversity. "The law of association examines the nature of relations and recognizes that order can be not just upset, but totally destroyed, if connections are impoverished. Diversity is the key to success."[3]

So, according to Watson's ecological insights, as a species we need to learn to connect with our place, the Earth, both locally and globally. We need to learn to live in balance with the limits of the planet, in our numbers and distribution. And we need to acknowledge diversity as the strength of the system without which we all become impoverished and stand a weakened chance of survival.

We also need to grow a culture that reflects these biological realities. As human beings, our cultural capacity is perhaps our greatest strength. As a species, we must move from selfish ego-centered behavior to behavior that is eco-centered. Community characterizes ecological law, and envisioning ourselves within the larger biological whole will have a profound impact on our culture.

In renewing the human community around honoring the Earth, we are embarking on the greatest enterprise that humanity has ever undertaken. This sacred journey—understanding ourselves as a collective species in balance with the larger circle of vibrant life—allows us to reinvent ourselves in harmony with the majestic web of nature.

The very act of restoring the Earth will restore our spirit. Restoration ecologists repairing damaged ecosystems have made hopeful discoveries. When people work together to regenerate the land, the land rebounds with vitality. Species reappear that were believed lost, but were only dormant. Nature is strong and resilient, the life force more powerful and mysterious than we can imagine. How could we have thought we could destroy it? In harming the Earth, we harm ourselves above all. The environmental crisis is also our spiritual catharsis. In healing the Earth, we heal ourselves.

There are many who are depending on us to awaken from the nightmare we have created. As we learn to see the tears of the fishes choking in poisoned waters, and to hear the cries carried on the beating wings of birds bereft of their forest homes, we come to know that the spirit of the Earth is within us.

Some say that the souls of departed fish remain in the sea, that one need only sing the right songs to bring them back. The bioneers are hearing those songs. May we all learn to listen and sing the songs sweetly.

Resource Section

Chapter 1 (Water)

BIONEERS

Dr. John Todd, Ocean Arks International, 233 Hatchville Rd., East
Falmouth, MA 02536; 508-563-2792; fax: 508-563-2880; e-mail:
bjosephs@cape.com; web address: www.vsp.cape.com/~bjosephs.
Also available are the terrific *Annals of Earth* newsletters and several
books by Dr. Todd. Ocean Arks can advise on personal and commercial living machines.

Dr. Donald Hammer, Hammer Resources, Inc., P.O. Box 65, Norris, TN
37828; 615-494-0388; fax: (615)494-0479. Books by Dr. Hammer
below. Dr. Hammer consults on constructed wetlands for all types of
wastewater treatment.

ORGANIZATIONS

Ecological Design Society, P.O. Box 11645, Berkeley, CA 94712; e-mail:
ecodesign@igo.apc.org.

National Small Flows Clearinghouse, West Virginia Univ., P.O. Box
6064, Morgantown, WV 26506-6064; 800-624-8301. Offers top
newsletter on alternative wastewater systems for small communities.

River Network, P.O. Box 8787, Portland, OR 97207; 503-241-3506;
800-423-6747.

Trout Unlimited, 1500 Wilson Blvd., Ste. 310, Arlington, VA 22209;
703-522-0200.

Center for Marine Conservation, 1725 Desales St. NW, Ste. 600, Washington, DC; 202-429-5609.

PUBLICATIONS

Books

Constructed Wetlands for Wastewater Treatment and Creating Freshwater
Wetlands, D. A. Hammer, ed. (Chelsea, MI: Lewis Publishers, 1989).
Wetlands Systems in Water Pollution Control, D. A. Hammer and R. L.
Knight (Oxford, UK: Pergamon Press, 1990) and Constructed Wetlands in Water Pollution Control, P. C. Cooper and B. C. Findlater,
eds. (Oxford, UK: Pergamon Press, 1990).
Adopting a Stream, Steven Yates (Seattle, WA: Univ. of Washington
Press). Order from Univ. of Washington Press, P.O. Box 50096,
Seattle, WA 98145; 800-441-4115.
Cadillac Desert, Marc Reisner (New York: Penguin Books, 1993).
The Sea Around Us, Rachel L. Carson (New York: Oxford Univ. Press,
1991). Order from Oxford Univ. Press/Order Dept., 2001 Evans Rd.,
Cary, NC 27513; 800-451-7556.
Gray Water Use in The Landscape, Robert Kourik (Santa Rosa, CA:
Metamorphic Press, 1948). Order from Metamorphic Press, P.O. Box
1841, Santa Rosa, CA 95402; 707-874-2606.
The Drinking Water Book, Colin Ingram (Berkeley, CA: Ten Speed Press,
1991). Order from Ten Speed Press, P.O. Box 7123, Berkeley, CA 94707.
Island Press has many excellent water and technical environmental
books. Order from Island Press/Order Dept., Box 7, 24850 Fast Lane,
Covelo, CA 95428; 800-828-1302.

Journals and Magazines

Coastal Connection (quarterly), Center for Marine Conservation, Chesapeake Field Office, 306A Buckroe Ave., Hampton, VA 23664.
World Rivers Review, International Rivers Network, 1847 Berkeley Way,
Berkeley, CA 94703; 510-848-1155.
The Whole Earth Review, issue "Water Talks," no. 85, spring 1995. Contains an excellent resource list about watersheds and watershed activism. P.O. Box 3000, Denville, NJ 07834.

Newsletters

American Rivers Newsletter, 1025 Vermont Ave. NW, Ste. 720, Washington, DC 20003; 202-547-6900; fax: 202-543-6142.

National Wetlands Newsletter, Environmental Law Institute, 1616 P St. NW, Ste. 200, Washington, DC 20036; 202-328-5150.

Catalogs

Water Books Catalog, AgAccess, P.O. Box 2008, Davis, CA 95617; 916-756-7177.

Chapter 2 (Biodiversity)

BIONEERS

Dr. Vandana Shiva, A-60, Hauz Khas, New Delhi, 110016, India; fax: 91-11-6856795. See below for listing of books by Dr. Shiva.

Third World Network, 87 Cantonment Rd., 10250, Penang, Malaysia. Distributes Shiva's books and coordinates Third World environmental networks.

ORGANIZATIONS

Conservation International, 1015 18th St. NW, Ste. 1000, Washington, DC 20036; 202-429-5660.

International Union for the Conservation of Nature, 1400 16th St. NW, Ste. 502, Washington DC 20036; 202-797-5454.

World Wildlife Fund, P.O. Box 4866, Hampden Station, Baltimore, MD 21211; 410-516-6951.

Botanical Preservation Corps, P.O. Box 1368, Sebastopol, CA 95473.

Native Seed/SEARCH, 2059 N. Campbell Ave., #325, Tucson, AZ 85719. Works to preserve Southwest Native American seed and farming heritage.

Allies, P.O. Box 2422, Sebastopol, CA 95473. Preserves and disseminates rare medicinal and entheonenic plants and seeds.

Rainforest Action Network (RAN), 4501 Sansome, Ste. 700, San Francisco, CA 94111; 800-989-7246. Offers *Amazonia*, the best citizen's guide to almost every organization in every nation trying to assist Amazonia.

Woodworkers Alliance for Rainforest Protection, P.O. Box 133, Coos Bay, OR 97420. Information and alternatives on sustainable wood product supplies.

Rainforest Alliance, 65 Bleecker St., 6th Floor, New York, NY 10012; 212-677-1900. Offers the Smart Wood project with timber certification, a medicinal plant project, fellowships, and education.

Center for Plant Conservation, Missouri Botanical Gardens, P.O. Box 299, St. Louis, MO 63166; 314-577-9450. A remarkable archive of rare and endangered plants. They save seeds and cuttings, keep track of which botanical gardens or arboreta have which plants, maintain a rare plants database, and supply speakers, brochures, slide shows, and directories.

New England Wildflower Society, Garden-in-the-Woods, Framingham, MA 01701. The model bioregional society. Propagated plants, great seeds, and catalogs.

North Carolina Botanical Garden, Box 3375, Totten Center, Univ. of North Carolina, Chapel Hill, NC 27514. Propagated seeds for Southeastern plants.

National Wildflower Research Center, 2600 FM 973 North, Austin TX 78725. Their *Wildflower Handbook* offers the most comprehensive list of concerned groups.

Conservation Monitoring Centre, The Herbarium, Royal Botanic Gardens, Kew, Richmond, Surrey TW93AB, England.

Foundation on Economic Trends, 1660 L St. NW, Ste. 216, Washington, DC 20036; 202-466-2823. Jeremy Rifkin's organization is known for blistering biotech critiques and related public-interest lawsuits.

Food and Water, RR1, Box 680, Walden, VT 05873; 800-EAT-SAFE.

International Center for Technology Assessment, 310 D St. NE, Washington DC 20002; 202-547-9359.

PUBLICATIONS

Books

Staying Alive: Women, Ecology & Development, Vandana Shiva (London: Zed Publishers, 1989).

Eco-Feminism, Maria Mils and Vandana Shiva (London: Zed Publishers, 1993).

Monocultures of the Mind: Biodiversity, Biotechnology and the Third World, Vandana Shiva (Penang, Malaysia: Third World Network, 1993).

The Violence of the Green Revolution: Third World Agriculture, Ecology and Politics, Vandana Shiva (Goa, India: The Other India Press, 1992).

Close to Home: Women Reconnect Ecology, Health, & Development World-wide, Vandana Shiva, ed. (Philadelphia: New Society Publishers, 1994).

Bio-Piracy: The Plunder of Nature and Knowledge, Vandana Shiva (Boston: South End Press, 1997).

Seeds of Change: The Living Treasure, Kenny Ausubel (San Francisco: HarperSanFrancisco, 1994).

Enduring Seeds, Gary Paul Nabhan (New York: North Point Press/Putnam Publishing, 1989).

Seed to Seed, Susanne Ashworth (Decorah, IA: Seed Savers Exchange, 1991). Seed-saving instructions.

The Diversity of Life, Edward O. Wilson (Cambridge, MA: Harvard Univ. Press, 1992). If you read only one biodiversity book, this may be it.

The Last Rain Forests: A World Conservation Atlas, Mark Collins, ed. (Oxford, UK: Oxford Univ. Press, 1990).

The Gardener's Guide to Plant Conservation, Nina T. Marshall (Baltimore, MD: World Wildlife Fund, 1993).

Birds in Jeopardy, Paul Ehrlich, David Dobkin, and Darryl Wheye (Stanford, CA: Stanford Univ. Press, 1992).

The Gene Hunters: Biotechnology and the Scramble for Seeds, Calestous Juma (Princeton, NJ: Princeton Univ. Press, 1990).

Perils Amidst the Promise: Ecological Risk of Transgenic Crops in a Global Market, Jane Rissler and Margaret Mellon (Cambridge, MA: Union of Concerned Scientists, 1993).

Biotechnology's Bitter Harvest: Herbicide-Tolerant Crops and the Threat to Sustainable Agriculture, Rebecca Goldburg, Jane Rissler, et al. (Vienna, VA: National Wildlife Federation, 1990).

Whole Earth Millennium Catalog (San Francisco: HarperSanFrancisco, 1995).

Newsletters

Vanishing Rainforest Education Kit, with video from World Wildlife Fund. Good place for teachers of grades two to six to start. Order from World Wildlife Fund, P.O. Box 4866, Baltimore, MD 21211; 410-516-6951.

Chapter 3 (Cultural Diversity)

BIONEERS

Francisco Alarcón, Univ. of California, Davis, Dept. of Spanish, Davis, CA 95616-8702; 916-752-1022; secretary: 916-752-0835; fax: 916-752-2184; e-mail: fjalarcon@ucdavis.edu. See below for books.

ORGANIZATIONS

Anti-Slavery Society, 180 Brixton Rd, London SW96AT England; 44-71-587-0573. The world's oldest human rights organization.

Native American Public Broadcasting Consortium, P.O. Box 83111, Lincoln, NE 68501; 402-472-3522. Large media catalog.

Native American Rights Fund, 1506 Broadway, Boulder, CO, 80302-6296; 303-447-8760.

Cultural Survival, 215 First St., Cambridge, MA 02138; 617-621-3818.

The Ladakh Project, P.O. Box 9475, Berkeley CA 94709.

South and Meso-American Indian Information Center, P.O. Box 28703, Oakland, CA 94604; 510-524-0795. News source and networking group for the indigenous peoples of South and Central America.

Coordinating Body for Indigenous People's Organization, Calle Alemania 836 y Mariana de Jesus, Quito, Ecuador. Alliance of Indian organizations in Brazil, Peru, Ecuador, Colombia doing information dissemination.

Indian Law Resource Center, 601 E St. SE, Washington DC 20003; 202-547-2800, 406-449-2031. Free legal help in major cases involving Indian rights. Helps North American natives with the management of trust funds; the reform of discriminatory laws; works toward sovereignty in Central and South America.

American Indian Law Alliance, 404 Lafayette St., New York, NY 10003; 212-598-0100.

PUBLICATIONS

Books

Snake Poems: An Aztec Invocation, Francisco Alarcón (San Francisco: Chronicle Books, 1992).

Aztec Sorcerers in Seventeenth Century Mexico: The Treatise on Superstitions by Hernando Ruiz de Alarcón, Publication No. 7 1982 (Albany, NY: State Univ. of New York at Albany, Institute for Mesoamerican Studies, 1982).

State of the Peoples: A Global Human Rights Report on Societies in Danger, Marc Miller, ed. (Boston: Beacon Press, 1993).

Millennium: Tribal Wisdom and the Modern World, David Maybury-Lewis (New York: Viking, 1992).

American Indian Myths and Legends, Richard Erdoes and Alfonso Ortiz (New York: Pantheon Books, 1985).

Black Elk Speaks, John G. Neihardt (Lincoln: Univ. of Nebraska Press, 1979).

House Made of Dawn, N. Scott Momaday (New York: Harper Collins, 1977).

The Conquest of Paradise, Kirkpatrick Sale (New York: Plume Books, 1990).

Ancient Futures, Helena Norberg-Hodge (San Francisco: Sierra Club Books, 1991).

The Law of the Mother : Protecting Indigenous Peoples in Protected Areas, Elizabeth Kempf, ed. (San Francisco: Sierra Club Books, 1993).

Treatise on the Heathen Superstitions That Today Live Among the Indians Native to This New Spain, 1629, Richard J. Andrews and Ross Hassig (Norman: Univ. of Oklahoma Press, 1984).

The Mythology of Mexico and Central America, John Bierhost (New York: William Morrow, 1992).

The Sacred Path: Spells, Prayers and Power Songs of the American Indians (New York: William Morrow, 1984).

2-Rabbit 7-Wind: Poems From Ancient Mexico Retold from Nahuatl Texts (New York: Viking, 1971).

Fifteen Poets of the Aztec World, Miguel Leon-Portilla (Norman: Univ. of Oklahoma Press, 1992).

Native Mesoamerican Spirituality (New York: Paulist Press, 1980).

The Broken Spears: Aztec Account of the Conquest of Mexico (Boston: Beacon Press, 1966).

Aztec Thought and Culture: A Study of the Ancient Nahuatl Mind (Norman: Univ. of Oklahoma Press, 1963).

245

Indian Givers: How the Indians of the Americas Transformed the World, Jack Weatherford (New York: Ballantine Books, 1987).

Chapter 4 (Green Medicine)

BIONEERS

Kathleen Harrison, Botanical Dimensions, P.O. Box 807, Occidental, CA 95465; 707-874-1531; fax: 707-874-2336; e-mail: botanize@wco.com. Harrison leads expeditions to BD's Hawaii garden and the Sachamama garden in Peru. She publishes an intermittent newsletter *PlantWise*.

Dr. Steven King, Shaman Pharmaceuticals, 213 E. Grand, South San Francisco, CA 94080; 415-952-7070.

Katy Moran, Executive Director, Healing Forest Conservancy, 3521 "S" St. NW, Washington, DC 20007.

ORGANIZATIONS

Conservation International, 1015 18th St. NW, Ste. 1000, Washington, DC 20036; 202-429-5660.

Institute of Economic Botany of the New York Botanical Garden, New York Botanical Garden, Dr. Michael Balick, Bronx NY 10458; 212-220-8777.

Michael McColm, Fundacion Jatun Sacha, Casilla 17-12-867, Avenida Rio Coca 1734, Quito, Ecuador; 011-593-441-592.

Genetic Resources Communications Systems, Inc., Deborah G. Strauss, Diversity, 4905 Del Ray Ave., Ste. 401, Bethesda, MD 20814; 301-907-9350, fax: 301-907-9328; e-mail: kingsbury@faculty.law.duke.edu.

Indigenous Peoples Biodiversity Network, Alejandro Argumento, Cultural Survival Canada, 200 Isabella St., Ste. 304, Ottawa ON, Canada K1S17; 613-237-5361; fax: 613-237-1547, e-mail: ipbn@web.apc.org or csc@web.apc.org.

PUBLICATIONS
Books

Tales of a Shaman's Apprentice, Mark Plotkin (New York: Viking, 1993).

Conservation of Medicinal Plants, V. Heywood and H. Synge (New York: Cambridge Univ. Press, 1991).

The Healing Forest: Medicinal and Toxic Plants of the Northwest Amazonia, R. Raffauf and R. Schultes (Portland, OR: Dioscorides Press, 1990).

Plants and People of the Golden Triangle: Ethnobotany of the Hill Tribes of Northern Thailand, E. Anderson (Portland, OR: Dioscorides Press, 1993).

Handbook of African Medicinal Plants, M. Iwu (Boca Raton, FL: CRC Press, 1993).

Amazonian Ethnobotanical Dictionary, J. Duke and R. Vasquez (Boca Raton, FL: CRC Press, 1993).

The Ethnobotany Book, G. Martin (New York: Chapman & Hall, 1995).

Highlands, Plants and Polynesians: Introduction to Polynesian Ethnobotany, P. Cox and S. Bannack (Portland, OR: Dioscorides Press, 1991).

Indigenous Knowledge and Intellectual Property Rights, S. Brish, D. Stablinsky, eds. (Covelo, CA: Island Press, 1995).

Journals and Articles

HerbalGram, The Journal of the American Botanical Council and the Herb Research Foundation, P.O. Box 201660, Austin, TX 78720. Superlative nonprofit magazine on plant medicine and herbs.

"From Shaman to Human Clinical Trials: The Role of Industry in Ethnobotany, Conservation and Community Reciprocity," in *Ciba Foundation Symposium No. 185,* Steven King and M. Tempesta (New York: John Wiley, 1994).

"Tropical Plants as a Source of New Pharmaceuticals," in *Pharmaceutical Manufacturing International, 1994,* M. Tempesta and Steven King (London: Sterling Publications, Ltd., 1994).

"Ethnobotany as a Source for New Drugs," in *Annual Reports in Medicinal Chemistry-29,* M. Tempesta and Steven King (Orlando, FL: Academic Press, 1994).

VIDEO

Hoxsey: How Healing Becomes a Crime, Realidad Productions/Kenny Ausubel, P.O. Box 1644, Santa Fe, NM 87504. Award-winning film by Ausubel on the civil war between organized and alternative medicine over a famous herbal remedy.

Chapter 5 ("Agri-culture")

BIONEERS

Fred Kirshenmann, Farm Verified Organic Inc., RR 1, Box 40A, Medina, ND 58467; 701-486-3578; fax: 701-486-3580.

Anna Edey, Solviva Solar-Dynamic Bio-Benign Design, RFD 1 Box 582, Vineyard Haven, MA 02568; 508-693-3341.

ORGANIZATIONS

Northern Plains Sustainable Agriculture Society, RR 1, Box 34, Fullerton, ND 58441.

Farm Aid, 334 Broadway, #5, Cambridge, MA 02139.

ATTRA (Appropriate Technology Transfer for Rural Areas), P.O. Box 3657, Fayetteville, AR 72702; 800-346-9140.

The Bio-Dynamic Farming & Gardening Association, P.O. Box 550, Kimberton, PA 19442; 800-516-7797, 610-935-7797.

John Jeavons Ecology Action of the Midpeninsula, 5798 Ridgewood Rd., Willits, CA 95490; 707-459-0150. Great resource for biointensive farming practices and mini-farms; author of classic *How to Grow More Vegetables Than You Ever Thought Possible on Less Land Than You Can Imagine* (see below).

Michael Fields Agricultural Institute, W. 2493 County Rd. ES, East Troy, WI 53120; 414-642-3303.

International Federation of Organic Agriculture Movements, Okozentrum Imsbach, D66636 Tholey-Theley, Germany; 49-6853-5190, fax: 49-6853-3011.

American Farmland Trust, 1920 N. St. NW, Ste. 400, Washington, DC 20036; 202-659-5170.

Land Institute, 2440 E. Water Well Rd., Salina, KS 67401; 913-823-5376; fax: 913-823-8728.

Organic Farming Research Foundation, Box 440, Santa Cruz, CA 95061; 408-426-6606.

Community Alliance With Family Farmers, P.O. Box 363, Davis, CA 95617; 916-756-8518.

California Certified Organic Farmers, 1115 Mission St., Santa Cruz, CA 95060; 408-423-2263; fax: 408-423-4528.

Oregon Tilth, P.O. Box 218, Tualatin, OR 97062; 503-691-9810.

Organic Crop Improvement Association, 1001 Y St., Ste. B, Lincon, NE 68508-1172; 402-477-2323, fax: 402-477-4325.

Organic Growers and Buyers Association, 7362 University Ave, Ste. 208, Fridley, MN 55432.

Organic Foods Production Association of North America, P.O. Box 1078, Greenfield, MA 01302; 413-774-7511, fax: 413-774-6432.

Independent Organic Inspectors Association, Rte. 3, Box 162-C, Winona, MN 55987; 507-454-8310.

Agricultural Library Forum, National Agricultural Library, 10301 Baltimore Ave., Beltsville, MD 20705; 301-504-5755. Largest agricultural library in the world.

PUBLICATIONS

Books

An Agriculture Testament, Sir Albert Howard (Oxford, UK: Oxford Univ. Press, 1943).

Agriculture, Lectures in 1924, Rudolf Steiner (Kimbertion, PA: Biodynamic Farming and Gardening Association, 1993).

Gardening by Mail, Barbara Barton (Boston: Houghton Mifflin, 1994).

The Unsettling of America, Wendell Berry (San Francisco: Sierra Club Books, 1986).

How to Grow More Vegetables Than You Ever Thought Possible on Less Land Than You Can Imagine, rev. ed., John Jeavons (Berkeley, CA: Ten Speed Press, 1995).

Permaculture, Bill Mollison (Australia: Tagari Publications, 1988).

Saving the Farm, American Farmland Trust (Davis, CA: American Farmland Trust, 1990). Order from American Farmland Trust Western Office, 1949 5th St., Ste. 101, Davis, CA 95616; 916-753-1073.

Start With the Soil, Grace Gershuny (Emmaus, PA: Rodale Press, 1993).

Beyond Beef, Jeremy Rifkin (New York: Plume Books, Penguin, 1993).

From the Good Earth: A Celebration of Growing Food Around the World, Michael Abelman (New York: Harry N. Abrams, 1993). Order from 100 5th Ave., New York, NY 10011; 800-345-1359.

Fertile Soil: A Grower's Guide to Organic & Inorganic Fertilizers, Robert Parnes (Davis, CA: AgAccess, 1990).

Directories and Catalogs

Healthy Harvest and Organic Market Guide, New Jersey Region, AgAccess, P.O. Box 2008, Davis, CA 95617; organic producers; 916-756-7177.

Farming in Nature's Image, Judith D. Soule and Jon K. Piper (Covelo, CA: Island Press, 1992). Order from Island Press, Box 7, Covelo, CA 95428; 800-828-1302.

Directory of Organic Agriculture, Box 116, Collingwood, Ontario, L9Y 3Z2, Canada; 705-444-0923; fax: 705-444-0380.

AgAccess (big catalog of sustainable ag books), P.O. Box 2008, Davis, CA 95617.

Magazines

Acres USA, P.O. Box 8800, Metairie, LA 70011. Great newspaper on sustainable agriculture.

VIDEO

My Father's Garden, Miranda Productions, P.O. Box 4624, Boulder, CO 80306; 303-546-0880; fax: 303-546-0990; e-mail: miranda@igc.apc.org. Fine documentary with segment on Fred Kirshenmann.

GARDENING AND SEED CATALOGS

Gardener's Supply, 128 Intervale Rd., Burlington, VT 05401; 800-863-1700.

Gardens Alive!, 5100 Schenley Place, Lawrenceburg, IN 47025; 812-537-8650.

Harmony Farm Supply, P.O. Box 460, Graton, CA 95444; 707-823-9125, 707-823-1734.

Peaceful Valley Farm Supply, P.O. Box 2209, Grass Valley, CA 95945; 916-272-4769.

Necessary Trading Co., One Nature's Way, New Castle, VA 24127-0305.

Seeds of Change, P.O. Box 15700, Santa Fe, NM 87506-5700; 505-438-8080.

The Cook's Garden, P.O. Box 65, Londonderry, VT 05148.

J. L. Hudson Seedsman, P.O. Box 1058, Redwood City, CA 94064.

Allies, P.O. Box 2422, Sebastopol, CA 95473.

The Redwood City Seed Company, Craig Dremann, Proprietor, P.O. Box 361, Redwood City, CA 94064.

Native Seed/SEARCH, 2509 N. Campbell Ave., Tucson, AZ 85719.
Seed Savers Exchange, Rural Route 3, Box 239, Decorah, IA 52101.

SUPPLIERS OF BENEFICIAL ORGANISMS

Rincon Vitova Insectaries, 805-543-5407.
Charles D. Hunter, EPA/ Dept. of Pesticide Regulation, 1020 "N" St.,
Room 161, Sacramento, CA 95814; 916-324-4100.

Chapter 6 ("Eco-nomics")

BIONEERS

Joshua Mailman, Sirius Business, 150 E. 58 St., 14th Floor, New York,
NY 10155; 212-421-3131; fax: 212-421-3163.
Jason Clay, World Wildlife Fund, 1250 24th St. NW, Washington DC
20037-1175; 202-778-9691; fax: 202-293-9211; or 2253 North
Upton St., Arlington, VA 22207; fax: 703-524-0092.

ORGANIZATIONS

Social Venture Network/U.S., P.O. Box 29221, San Francisco, CA
94129; 415-561-6501; fax: 415-771-0535.
Social Venture Network/Europe, P.O. Box 973, 1000 AX Amsterdam,
The Netherlands; 31 20 535 3250.
Business for Social Responsibility, 609 Mission St., 2nd floor, San Fran-
cisco, CA 94105; 415-537-0888; website: www.bsr.org.
Investor Circle, 3220 Sacramento St., San Francisco, CA 94115;
415-929-4910; fax: 415-929-4915; e-mail: icircle@aol.com.
Council on Economic Priorities, 30 Irving Place, New York, NY 10003;
800-729-4237.
Co-Op America, 1612 K St. NW, #600, Washington, DC 20006;
202-872-5307.
CERES (Coalition for Environmentally Responsible Economies),
711 Atlantic Ave., Boston, MA 02111; 617-451-0927.
Public Citizen, 2000 "P" St. NW, Washington, DC 20036; 202-833-3000.
Ralph Nader's public interest group which monitors corporations.

International Institute for Sustainable Development, 161 Portage Ave., E6th Floor, Winnipeg, Manitoba, Canada, R3B 0Y4; 204-958-7700.

Human Rights Watch, 485 Fifth Ave., New York, NY 10017; 212-972-8400; fax: 212-972-0905.

Southwest Network for Environmental and Economic Justice, P.O. Box 7399, Albuquerque, NM 87194-7399; 505-242-0416; fax: 505-242-5606.

Social Investment Forum, P.O. Box 57216, Washington, DC 20037; 202-833-5522.

Kinder, Lydenberg and Domini, 120 Mt. Auburn St., Cambridge, MA 02138; 617-547-7479.

SOCIALLY RESPONSIBLE INVESTMENT FUNDS/ BANKS/MANAGERS

Progressive Asset Management, Inc. (PAM), 1814 Franklin St., #710, Oakland, CA 94612; 510-834-3722.

South Shore Bank, 71st and Jeffery Blvd., Chicago, IL 60649; 800-669-7725.

Vermont National Bank, P.O. Box 804, Brattleboro, VT 05302; 800-SRB-FUND.

Cascadia Revolving Fund, 157 Yesler Way, #414, Seattle, WA 98104; 206-447-9226.

Community Capital Bank, 111 Livingston St., Brooklyn, NY 11201; 718-802-1212.

Alternatives Federal Credit Union, 301 W. State St., Ithaca, NY 14850; 607-273-4666.

Calvert Group, 4550 Montgomery Ave., Bethesda, MD 20814; 800-368-2748.

Domini Social Equity Fund, 6 St. James Ave., Boston, MA 02116; 800-762-6814.

Green Century Funds, 29 Temple Place, Boston, MA 02111; 800-93-GREEN.

New Alternatives Funds, Inc., 150 Broadhollow Rd., Ste. 306, Melville, NY 11747; 516-423-7373.

Parnassus Fund, 244 California St., San Francisco, CA 94111; 800-999-3505.

Pax World Fund, 224 State St., Portsmouth, NH 03801; 800-767-1729.
Citizens Trust, One Harbour Place, #225, Portsmouth, NH 03801;
800-223-7010.
Harrington Investments, P.O. Box 6108, Napa, CA 94581-1108;
800-788-0154.

PUBLICATIONS

Books

Generating Income and Conserving Resources: Twenty Lessons From the Field, Jason Clay (Washington DC: World Wildlife Fund, 1996).
Market Potentials for Redressing the Environmental Impact of Wild, Captured and Pond Produced Shrimp, Jason Clay (Washington, DC: World Wildlife Fund, 1997).
The Ecology of Commerce, Paul Hawken (New York: HarperCollins, 1993).
The Next Economy, Paul Hawken (New York: Ballantine, 1983).
For the Common Good, Herman E. Daly and John B. Cobb, Jr. (Boston, MA: Beacon Press, 1991).
How Much Is Enough?, Alan Durning (New York: W. W. Norton, 1992); great critique of consumption.
Trade and the Environment, Durwood Zaelke, Paul Orbuch, and Robert F. Housmann, eds. (Covelo, CA: Island Press, 1993).
How the West Grew Rich, Nathan Rosenberg and L. E. Birdzell, Jr. (New York: Basic Books, 1987).
Rival Views of Market Society, Albert O. Hirschman (Cambridge, MA: Harvard Univ. Press, 1992).
The End of Equality, Mickey Kaus (New York: HarperCollins, 1993).
The Work of Nations, Robert B. Reich (New York: Random House, 1992).
Confronting Environmental Racism, Robert D. Bullard, ed. (Boston: South End Press, 1993). Order from South End Press, 116 St. Botolph St., Boston, MA 02115; 800-533-8478.
Toward a History of Needs, Ivan Illich (Berkeley, CA: Heyday Books, 1978). Order from P.O. Box 9145, Berkeley, CA 94709; 510-549-3564.

Paradigms in Progress, Hazel Henderson (Indianapolis, IN: Knowledge Systems, 1992). Order from Knowledge Systems, 7777 W. Morris St., Indianapolis, IN 46231; 800-999-8517.

Small Is Beautiful, E. F. Schumacher (New York: Harper & Row Publishers, 1973).

Beyond the Limits, Donella H. Meadows, Dennis L. Meadows, and Jorgen Randers (Lebanon, NH: Chelsea Green Publishing, 1992). Order from Chelsea Green Publishing Co., 10 Water St., Lebanon, NH 03766; 800-639-4099.

The Global Citizen, Donella H. Meadows (Covelo, CA: Island Press, 1991). Order from Island Press, Box 7, Covelo, CA 95428; 800-828-1302.

Toxic Waste and Race in the United States, United Church of Christ Commission for Racial Justice, 700 Prospect Ave., Cleveland, OH 44115.

Magazines and Newsletters

Good Money; 800-535-3551; website: www.goodmoney.com.

The Greenmoney Journal, W. 608 Glass Ave., Spokane, WA 99205; 509-328-1741; website: www.greenmoney.com.

Race, Poverty and the Environment, Urban Habitat Program, Earth Island Institute, 300 Broadway, Ste. 28, San Francisco, CA 94133; 415-788-3666.

Bank Check, Juliette Majot, ed., Bank Check, c/o International Rivers Network, 1847 Berkeley Way, Berkeley, CA 94703;510-848-1155.

The Green Business Letter, 1519 Connecticut Ave. NW, Washington, DC 20036; 800-955-GREEN.

Chapter 7 (Redesigning Society)

BIONEERS

Monika Griefahn, Niedersächsische Umweltministerin, 30041 Hannover, Postfach 4107, Archivstrasse 2, Germany; phone: 0511-120-3301; fax: 0511-120-3199.

William McDonough and Associates, 410 E. Water St., Charlottesville, VA 22902; 804-979-1111, fax: 804-979-1112.

William McDonough, Dean, School of Architecture, Edward E. Elson
Endowed Chair, Campbell Hall, Univ. of Virginia, Charlottesville,
VA 22903; 804-924-7019.
McDonough Braungart Design Chemistry, 410 East Water St., Ste. 1100,
Charlottesville, VA 22902; 804-295-1111; fax: 804-295-1500;
e-mail: mbdc@mbdc.com.

ORGANIZATIONS

Rocky Mountain Institute, Amory and Hunter Lovins, 1739 Snowmass
Creek Rd., Snowmass, CO 81654-9199. The final word on appropri-
ate technology and energy efficiency.
The Center for Maximum Potential Building Systems, 8604 F.M. 969,
Austin, TX 78724; 513-928-4786.

PUBLICATIONS

Books

Living by Design, Sim Van Der Ryn and Stuart Cowan (Covelo, CA:
Island Press, 1995). Terrific work on state-of-the art eco-design.
The Natural House Book, David Pearson (New York: Simon & Schuster,
1989).
Energy Efficiency and Human Activity, Lee Schipper and Stephen Meyers
(New York: Cambridge Univ. Press, 1992).
The Green Consumer, Joel Makower, John Elkington, and Julia Hailes
(New York: Penguin Books, 1993).
Guide to Resource Efficient Building Elements (Missoula, MT: Center
for Resourceful Building Technology, 1994). Order from CRBT,
P.O. Box 100, Missoula, MT 59806; 406-549-7678.
Sourcebook for Sustainable Design (Boston: Boston Society of Architects,
1992). Order from Boston Society of Architects, 52 Broad St.,
Boston, MA 02109; 617-951-1433, ext. 221.
How Buildings Learn, Stewart Brand (New York: Viking, 1994).
Design With Nature, Ian L. McHarg (New York: John Wiley & Sons,
1991).
Beyond 40 Percent, Institute for Local Self-Reliance Staff (Covelo, CA:
Island Press, 1991).

Business Recycling Manual, INFORM and Recourse Systems Staff (New York: INFORM, 1991). Order from INFORM, Inc., 120 Wall St., 16th Floor, New York, NY 10005-4001; 212-361-2400.

Sustaining the Earth, Debra Dadd-Redalia (New York: Hearst Books, 1994).

SUSTAINABLE PRODUCTS AND CATALOGS

Livos Non-Toxic Home Products, 1365 Rufina Circle, Santa Fe, NM 87501; 505-438-3448.

Hendericksen Natural Flooring and Interiors, P.O. Box 1677, Sebastopol, CA 95473; 707-824-0914.

Real Goods, 966 Mazzoni St., Ukiah, CA 95482; 800-762-7325. Real Goods is a retail supplier of innumerable appropriate technologies, including solar and wind.

Auro Natural Plant Chemistry Catalog, P.O. Box 857, Davis, CA 95617-0857; 916-753-3104.

Resource Conservation Technology, Inc., 2633 N. Calvert St., Baltimore, MD 21218; 410-366-1146.

Chapter 8 (Restoring Spirit)

BIONEERS

Jennifer Greene, Water Research Institute and Waterforms Inc., P.O. Box 930, Blue Hill, ME 04614; 207-374-2384; fax: 207-374-2383. Flow forms, consultations using drop-picture photography, water remediation, and educational lectures with fantastic slide shows on water.

John Perkins, Dream Change Coalition and Prydwen, Inc., P.O. Box 31357, Palm Beach Gardens, FL 33420; 561-622-6064; e-mail: http://www.dreamchange.org/dreamchange. Also, Buy Back Polluted Air (POLE); Perkins leads workshops and tours to shamanic retreat centers in Latin America.

ORGANIZATIONS

KATALYSIS, North/South Development Partnerships, 1331 N. Commerce St., Stockton, CA 95202; 209-943-6165.

American Indian Religious Freedom Coalition, Native American
Rights Fund, 1506 Broadway, Boulder, CO 80302-6296;
303-447-8760.

The Wildlands Project, 1955 W. Grant Rd., Ste. 148-A, Tucson, AZ,
85745; 520-884-0875.

Sea Shepherd Conservation Society, 3107-A Washington Blvd., Marina
Del Rey, CA 90292; 310-301-7325.

Greenpeace, 1436 U St. NW, Washington, DC 20009; 800-326-0959.

Sierra Club, 85 2nd St., 2nd Floor, San Francisco, CA 94105; 415-776-
2211; fax: 415-776-0350.

Audubon Society, 613 Riversville Rd., Greenwich, CT 06831; 203-869-
2017; fax: 203-869-4437.

The Wilderness Society, 900 17th St. NW, Washington DC 20006-
2596; 202-429-2637/3952.

Association of Forest Service Employees for Environmental Ethics
(AFSEE), P.O. Box 11615, Eugene, OR 97440; 541-484-2692.

Friends of the Earth International, 218 D St. SE, Washington, DC
20003; 202-783-7400.

Earth Island Institute, 300 Broadway, Ste. 28, San Francisco, CA 94133;
415-788-3666.

PUBLICATIONS
Books

*The World Is as You Dream It: Shamanic Teachings From the Amazon
and the Andes*, John Perkins (Rochester, VT: Destiny Books, 1994).
Order from Inner Traditions International, 1 Park St., Rochester,
VT 05767; 802-767-3174.

Psychonavigation: Techniques for Travel Beyond Time, John Perkins
(Rochester, VT: Destiny Books, 1990).

*The Stress-Free Habit: Powerful Techniques for Health and Longevity From
the Andes, Yucatan, and Far East*, John Perkins (Rochester, VT:
Healing Arts Press, 1989).

Wisdom of the Elders: Sacred Native Stories of Nature, David Suzuki and
Peter Knudtson (New York: Bantam, 1992).

Wizard of the Upper Amazon, F. Bruce Lamb and Manuel Cordova-Rios
(Berkeley, CA: North Atlantic Books, 1975).

Rio Tigre and Beyond: The Amazon Jungle Medicine of Manuel Cordova-Rios, F. Bruce Lamb (Berkeley, CA: North Atlantic Books, 1985).

Ayahuasca Visions, Pablo Amaringo and Luis Eduardo Luna (Berkeley, CA: North Atlantic Press, 1992).

Life in Moving Fluids, Steven Vogel (Princeton, NJ: Princeton Univ. Press, 1941).

The Science and Romance of Selected Herbs used in Medicine and Religious Ceremony, Anthony K. Andoh (San Francisco: North Scale Institute, 1986). Order from North Scale Institute, 2205 Travel St., San Francisco, CA 94116; 415-759-9491.

Steps to an Ecology of Mind, Gregory Bateson (New York: Ballantine Books, 1972).

Mind and Nature, Gregory Bateson (New York: Bantam Books, 1979).

Journals and Catalogs

Wild Earth: The Wildness Project, P.O. Box 455, Richmond, VT 05477; 802-434-4077.

Rosetta Folios and Books, P.O. Box 4611, Berkeley, CA 94704.

Notes

PREFACE

1. Clive Ponting, *A Green History of the World* (New York: St. Martin's Press, 1992). This book contains the definitive documentation of human-induced ecological collapse throughout history.
2. Paul Hawken, personal conversation with the author, 1995.

INTRODUCTION

1. Associated Press, "U.N.: Time Running Out on Saving Environment," *The New Mexican*, May 8, 1992.
2. Robert D. Kaplan, "The Coming Anarchy," *The Atlantic Monthly*, Feb. 1994, pp. 57–58.
3. David T. Suzuki and Peter Knudtson, *Wisdom of the Elders* (New York: Bantam, 1992), p. 55.
4. James Lovelock, *Gaia: A New Look at Life on Earth* (Oxford, UK: Oxford Univ. Press, 1987).
5. *New York Times*, Science Times, Oct. 15, 1996; several articles in special section on microbes and their relationship to planetary ecology.
6. Edward O. Wilson, *Biophilia* (Cambridge, MA: Harvard Univ. Press, 1984), p. 35.
7. Lynn Margulis, *Early Life* (Boston: Science Books International, 1982), p. 138.
8. Peter Kropotkin, *Mutual Aid: A Factor of Evolution* (New York: Washington Square Press, 1916), pp. 1–2, p. v.
9. John G. Neihardt, *Black Elk Speaks* (New York: Washington Square Press, 1959), p. 1.

10. William K. Stevens, "Balance of Nature? What Balance Is That?," *New York Times*, Oct. 22, 1991.

CHAPTER ONE

1. Peter Warshall, "The Morality of Molecular Water," *Whole Earth Review*, no. 85, spring 1995.
2. Clive Ponting, *A Green History of the World* (New York: St. Martin's Press, 1992), p. 350.
3. Ibid.
4. Sandra Postel, *Last Oasis: Facing Water Scarcity* (New York: W. W. Norton, Worldwatch Environmental Alert Series, 1992), pp. 33–35.
5. Ibid., p. 73.
6. Clive Ponting, *A Green History of the World*, p. 351.
7. Annette McGivney, "Troubled Waters," *E magazine*, Sept.–Oct. 1993, vol. IV, no. 5, p. 35.
8. Clive Ponting, *A Green History of the World*, p. 370.
9. John Young, *State of the World* (New York: W. W. Norton, 1994), pp. 110–11.
10. Natural Resources Defense Council, "Testing the Waters," NRDC publication vol. III, June 1993, pp. 5–6.
11. World Resources Institute, *World Resources 1994–1995* (Oxford, UK: Oxford Univ. Press, 1993), p. 183.
12. Worldwatch Institute, "Net Loss: Fish, Jobs and the Marine Environment," *Worldwatch Paper 120*, 1994, p. 120.
13. Donald Hammer, ed., *Constructed Wetlands for Wastewater Treatment* (Chelsea, MO: Lewis Publishers, 1991).

CHAPTER TWO

1. Gary Paul Nabhan and Stephen L. Buchmann, "The Pollination Crisis," *The Sciences*, Jul.–Aug. 1996.
2. Edward O. Wilson, *The Diversity of Life* (Cambridge, MA: Harvard Univ. Press, 1992), p. 301.
3. Edith C. Stein, *The Environmental Sourcebook* (New York: Lyons and Burford, 1992), p. 119; covers these statistics.
4. William K. Stevens, "Green Revolution Is Not Enough, Study Finds," *New York Times*, Sept. 6, 1995.

CHAPTER THREE

1. James Brooke, "Tribes Get Right to 50 percent of Colombian Amazon," *New York Times*, Feb. 4, 1990.
2. Mark Plotkin, *Tales of a Shaman's Apprentice* (New York: Viking Penguin, 1993); extensive discussion throughout.
3. Clive Ponting, *A Green History of the World* (New York: St. Martin's Press, 1992), pp. 130–33.
4. Jason Clay, "Looking Back to Go Forward: Predicting and Preventing Human Rights Violations," Mark Miller, ed., *The State of the Peoples* (Boston: Beacon Press, 1993), p. 66.
5. Alan Durning, "Guardians of the Land: Indigenous Peoples and the Health of the Earth," *Worldwatch Institute Research Paper 112*, 1992, p. 17.
6. Francisco X. Alarcón, *Snake Poems: An Aztec Invocation* (San Francisco: Chronicle Books, 1992). Poems are reprinted here with the kind permission of the author and the publisher.

CHAPTER FOUR

1. James Sterngold, "Japan's Cedar Forests Are Man-Made Disaster," *New York Times*, January 17, 1995, B7.
2. John W. Diamond, "What Is Disease and What Are the Obstacles to Cure?," *Biological Therapy*, vol. 12, no. 4, 1994.
3. Edward O. Wilson, *The Diversity of Life* (Cambridge, MA: Harvard Univ. Press, 1992), pp. 282–85.
4. Ibid., p. 321.
5. Ryan J. Huxtable, "The Pharmacology of Extinction," *Journal of Ethnopharmacology*, 37, 1992, p. 1.

CHAPTER FIVE

1. National Academy of Sciences, *Lost Crops of the Incas: Little-Known Plants of the Andes With Promise for Worldwide Cultivation* (Washington, DC: National Academy Press, 1989); extensive discussion throughout.
2. Clive Ponting, *A Green History of the World* (New York: St. Martin's Press, 1992); extensive discussion throughout.

3. Dale Baretl, "Integrated Pest Management," Council on Environmental Quality, Washington, DC, 1980, p. VI; also, Robert Van der Bosch, *The Pesticide Conspiracy* (Garden City, NY: Anchor Books), p. 24.

4. Kenny Ausubel and John J. O'Connor, "Let's End Agribusiness as Usual," *New Age Journal*, Mar.–Apr. 1995, pp. 76–77.

5. Kent Whealy and Arllys Adelmann, *Seeds Savers Exchange, "The First Ten Years"* (Decorah, IA: Seed Saver Publications, 1986), p. 106.

6. Ibid., p. 102.

7. Marty Strange, *Family Farming — A New Economic Vision* (Lincoln: Univ. of Nebraska Press, 1988), p. 41.

8. Cary Fowler and Pat Mooney, *Shattering: Food, Politics and the Loss of Genetic Diversity* (Tucson: Univ. of Arizona Press, 1990), pp. 115–39.

9. Michael Greger, "More Serious Than AIDS," *Earth Island Journal*, summer 1996, pp. 27–33.

10. Andrew Kimbrell, speech at Bioneers Conference, 1996, audiotape from Bioneers Conference (Santa Fe, NM: Collective Heritage Institute, 1996).

11. Rudolf Steiner, *Agriculture: Spiritual Foundations for the Renewal of Agriculture* (Kimberton, PA: Biodynamic Farming and Gardening Association, 1993).

CHAPTER SIX

1. Sandra Postel, "Toward a New Eco-nomics," *Worldwatch* magazine, Sept.–Oct. 1990, p. 22.

2. Lester R. Brown, *State of the World 1994*, Worldwatch Institute, p. 34.

3. Lester R. Brown, *State of the World 1993*, Worldwatch Institute, p. 190.

4. Paul Hawken, *The Ecology of Commerce* (New York: Harper Business, 1993), pp. 92, 95.

5. "The Rich Are Getting Richer, etc., and It's Likely to Remain That Way," *New York Times*, Mar. 28, 1996.

6. Paul Hawken, *The Ecology of Commerce*, p. 135.

7. Jason DeParle, "Census Sees Falling Income and More Poor," *The New York Times*, Oct. 7, 1994.

8. Lester R. Brown, Christopher Flavin, and Sandra Postel, *Saving the Planet* (New York: W. W. Norton, 1991), p. 11.

9. Sandra Postel, "Toward a New Eco-nomics," pp. 20–28.
10. Christopher Flavin, "Storm Warnings: Climate Change Hits the Insurance Industry," *Worldwatch* magazine, Nov.–Dec. 1994, pp. 10–20.
11. Paul Hawken, *The Ecology of Commerce* (New York: Harper Business, 1993), p. 89.
12. Richard Grossman and Frank T. Adams, *Taking Care of Business* (Cambridge, MA: Charter, 1993); extensive discussion of corporate charters and revocation.
13. Lloyd Kurtz and Dan DiBartolomeo, "Socially Screened Portfolios: An Attribution Analysis of Relative Performance," *The Journal of Investing*, fall 1996, pp. 35–41.
14. Ronald Smothers, "Study Concludes that Environmental and Economic Health Are Compatible," *New York Times*, Oct. 19, 1994, p. A12.

CHAPTER SEVEN

1. Alan Durning, *How Much Is Enough?* (New York: W. W. Norton, 1992), pp. 55–56.
2. Ibid., p. 72.
3. World Resources Institute, *World Resources 1994–1995* (Oxford, UK: Oxford Univ. Press, 1993), p. 199.
4. Clive Ponting, *A Green History of the World* (New York: St. Martin's Press, 1992), p. 387.
5. Christopher Flavin and Nicholas Lenssen, "Powering the Future," Worldwatch Institute Paper 119, June 1994, p. 6.
6. World Resources Institute, *World Resources 1994–1995*, p. 219.
7. Paul Hawken, *The Ecology of Commerce*(New York: Harper Business, 1993), p. 73.
8. Lester R. Brown, Christopher Flavin, and Sandra Postel, *Saving the Planet* (New York: W. W. Norton, 1991), p. 11.

CHAPTER EIGHT

1. Alan Durning, *How Much Is Enough?* (New York: W. W. Norton, 1992), p. 52.
2. World Resources Institute, Allen Hammond, ed., *The 1994 Information Please Environmental Almanac* (New York: Houghton Mifflin, 1994), p. 80.

3. Alan Durning, *How Much Is Enough?*, p. 122.
4. Andrew Kimbrell, *The Human Body Shop* (New York: HarperCollins, 1993), pp. 176, 200.
5. *New York Times*, Jul. 18, 1995, p. A13.
6. Theodor Schwenk, *Sensitive Chaos: The Creation of Flowing Forms in Water and Air* (New York: Schocken Books, 1976), p. 9.
7. Ibid., p. 13.
8. Ibid., p. 20.
9. Ibid., p. 21.

EPILOGUE

1. Warren E. Leary, "Science Takes a Lesson From Nature, Imitating Abalone and Spider Silk," *New York Times*, Aug. 31, 1993, pp. B5–6.
2. Lyall Watson, *Dark Nature: A Natural History of Evil* (New York: HarperCollins, 1995), pp. 9, 24.
3. Ibid., p. 42.

Index

The Bioneers Conference

PRACTICAL SOLUTIONS FOR RESTORING THE EARTH

As human beings strain the limits of the natural world, we can no longer escape the knowledge that ecological collapse has been the hidden history behind the downfall of many major civilizations. Biology is indeed destiny, as the ultimate fate of our species rests upon our ability to live within the limits of the natural world. Today we are witnessing the widespread collapse of the basic life-support systems upon which all life depends.

Restoring the Earth is destined to become the central enterprise of the years ahead, and leading the effort is a growing movement of "bioneers," biological pioneers who are using nature to heal nature. The bioneers bring practical solutions for virtually all our crucial environmental problems. These working models hold keys to planetary survival which can be refined, replicated, and rapidly spread around the world. This optimistic bioneer vision is equipping individuals, groups, communities, and companies to act as a primary force in the fabric of restoration.

The bioneers are impassioned individuals who offer a positive vision of restoration. They are scientists, architects, educators, artists, spiritual leaders, farmers, gardeners, botanists, economists, entrepreneurs, activists, and public servants, pointing the way to a saner, more sustainable way of life through direct action. The bioneers herald a coming Age of Biology, one in which as human beings we weave our lives back into harmony with the natural world. The bioneers agree that imitation is the sincerest form of flattery, and they emulate natural systems to fashion benign technologies and just forms of social organization. Above all, they call for us to reconnect with the sanctity of life through the heart of nature.

273

Since 1990, the Bioneers Conference has brought together leading scientific and social visionaries with practical solutions for restoring the Earth. The Bioneers Conference has helped define the biological model and galvanize its many brilliant practitioners into a thriving culture. Their ideas and practices are so original and pragmatic as to uplift the spirit with their extraordinary creativity and ingenuity — to create a future environment of hope. The Bioneers Conference takes place annually in San Francisco. For conference information, contact: Collective Heritage Institute (CHI), 826 Camino de Monte Rey, #A-5, Santa Fe, NM 87501, 505-986-0366; www.bioneers.org.

A complete set of Bioneers audiotapes is available from 1990–1996. Contact CHI for listings of auditotapes.

ALSO BY KENNY AUSUBEL

"Hoxsey: How Healing Becomes a Crime"

This feature-length nonfiction film won the "Best Censored Stories" Award in 1990, associated with Bill Moyers, for its hard-hitting exposé of how organized medicine has obstructed and suppressed alternative treatments for cancer. The film follows the fascinating, larger-than-life story of Harry Hoxsey, an ex-coal miner who inherited his family's herbal cancer remedy and started a string of clinics across the United States. A colorful character who could have come from the pages of Mark Twain, Hoxsey battled the American Medical Association and federal government for 35 years, and was the first ever to win a libel judgment against the AMA after being labeled a "quack." The film was widely praised and strikes a firm blow for freedom of medical choice. It portrays the corporatization of medicine in the twentieth century, and foretells the renaissance of herbal medicine. Realidad Productions, 83 minutes, color, 1987. Coproduced by Kenny Ausubel and Catherine Salveson. Written and directed by Kenny Ausubel. Available for $29.95 plus $3.00 shipping from Realidad Productions, P.O. Box 1644, Santa Fe, NM 87504; 505-986-0366; fax: 505-986-1644.

CHELSEA GREEN

Sustainable living has many facets. Chelsea Green's celebration of the sustainable arts has led us to publish trend-setting books about organic gardening, solar electricity and renewable energy, innovative building techniques, regenerative forestry, local and bioregional democracy, and whole foods. The company's published works, while intensely practical, are also entertaining and inspirational, demonstrating that an ecological approach to life is consistent with producing beautiful, eloquent, and useful books, videos, and audio cassettes.

For more information about Chelsea Green, or to request a free catalog, call toll-free (800) 639-4099, or write to us at P.O. Box 428, White River Junction, Vermont 05001. Visit our Web site at www.chelseagreen.com.

Chelsea Green's titles include:

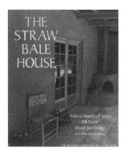

The Straw Bale House	The Neighborhood Forager	Believing Cassandra
The New Independent Home	The Apple Grower	Gaviotas: A Village to
The Natural House	The Flower Farmer	Reinvent the World
Serious Straw Bale	Breed Your Own	Who Owns the Sun?
The Beauty of	Vegetable Varieties	Global Spin:
Straw Bale Homes	Passport to Gardening	The Corporate Assault
The Resourceful Renovator	Keeping Food Fresh	on Environmentalism
Independent Builder	The Soul of Soil	Hemp Horizons
The Rammed Earth House	The New Organic Grower	A Patch of Eden
The Passive Solar House	Four-Season Harvest	A Place in the Sun
Wind Energy Basics	Solar Gardening	Beyond the Limits
Wind Power for Home &	Straight-Ahead Organic	The Man Who Planted Trees
Business	The Contrary Farmer	The Northern Forest
The Solar Living Sourcebook	The Co-op Cookbook	The New Settler Interviews
A Shelter Sketchbook	Whole Foods Companion	Loving and Leaving the
Mortgage-Free!	The Bread Builder	Good Life
Hammer. Nail. Wood.	Simple Food for the	Scott Nearing: The Making
Stone Circles	Good Life	of a Homesteader
Toil: Building Yourself	The Maple Sugar Book	Wise Words for the Good Life